Sustainable Energy for All
Transforming Commitments to Action
Lessons Learned and Actions for the Future

About The Centre

_____

The Centre for Science and Technology of the Non-Aligned and Other Developing Countries (NAM S&T Centre) is an inter-governmental organisation with a membership of 47 countries spread over Asia, Africa, Middle East and Latin America. Besides this, 12 S&T agencies and academic/research institutions of Bolivia, Botswana, Brazil, India, Nigeria and Turkey are the members of the S&T-Industry Network of the Centre. The Centre was set up in 1989 to promote South-South cooperation through mutually beneficial partnerships among scientists and technologists and scientific organisations in developing countries. It implements a variety of programmes including international workshops, meetings, roundtables, training courses and collaborative projects and brings out scientific publications, including a quarterly Newsletter. It is also implementing 6 Fellowship schemes, namely, NAM S&T Centre Research Fellowship, South Africa Mineral Processing Training Fellowship, Joint NAM S&T Centre – ICCBS Karachi Fellowship, Joint CSIR/CFTRI (Diamond Jubilee) - NAM S&T Centre Fellowship, Joint NAM S&T Centre – ZMT Bremen Fellowship and Research Training Fellowship for Developing Country Scientists (RTF-DCS) in Indian institutions. These activities provide, among others, the opportunity for scientist-to-scientist contact and interaction, training and expert assistance, familiarising the scientific community on the latest developments and techniques in the subject areas, and identification of technologies for transfer between member countries. The Centre has so far brought out 64 publications and has organised 95 international workshops and training programmes.

For further details, please visit www.namstct.org or write to the Director General, NAM S&T Centre, Core 6A, 2nd Floor, India Habitat Centre, Lodhi Road, New Delhi-110003, India (Phone: +91-11-24645134/24644974; Fax: +91-11-24644973; E-mail: namstcentre@gmail.com; namstct@bol.net.in).

Sustainable Energy for All

Transforming Commitments to Action

Lessons Learned and Actions for the Future

— Editors —

Moustafa Mohammed Eissa
R. Hari Kumar
Jyoti Prasad Painuly

CENTRE FOR SCIENCE & TECHNOLOGY OF THE
NON-ALIGNED AND OTHER DEVELOPING COUNTRIES
(NAM S&T CENTRE)

2016
DAYA PUBLISHING HOUSE®
A Division of
ASTRAL INTERNATIONAL PVT. LTD.
New Delhi – 110 002

Cataloging in Publication Data--DK
Courtesy: D.K. Agencies (P) Ltd. <docinfo@dkagencies.com>

Triennial International Workshop on 'Sustainable Energy for All: Transforming Commitments to Action' (3rd : 2014 : Bangalore, India)
Sustainable energy for all transforming commitments to action : lessons learned and actions for the future / editors, Moustafa Mohammed Eissa, R. Hari Kumar, Jyoti Prasad Painuly.
   pages cm
   Papers presented at the Third Triennial International Workshop on 'Sustainable Energy for All: Transforming Commitments to Action' organised by Centre for Science and Technology of the Non-Aligned and Other Developing Countries at Bangalore during 22-24 February 2014.

   ISBN 978-93-5130-876-8 (International Edition)

   1. Renewable energy sources--Developing countries--Congresses. 2. Renewable energy sources--Technological innovations--Developing countries--Congresses. 3. Energy policy--Developing countries--Congresses. I. Eissa, Moustafa Mohammed, editor. II. Harikumar, R., editor. III. Painuly, Jyoti Prasad, editor. IV. Centre for Science and Technology of the Non-Aligned and Other Developing Countries.organizer. V. Title.

   TJ807.9.D44T75 2014   DDC 333.794091724   23

Centre for Science and Technology of the Non-Aligned and Other Developing Countries (NAM S&T Centre)
Core-6A, 2nd Floor, India Habitat Centre, Lodhi Road,
New Delhi-110 003 (India)
Phone: +91-11-24644974, 24645134, Fax: +91-11-24644973
E-mail: namstct@gmail.com
Website: www.namstct.org

Published by	:	**Daya Publishing House®**
A Division of
Astral International Pvt. Ltd.
4760-61/23,. Ansari Road, Darya Ganj,
New Delhi - 110 002
Phone: 011-4354 9197, 2327 8134
E-mail: info@astralint.com
Website: www.astralint.com
Laser Typesetting	:	**Classic Computer Services**, Delhi - 110 035
Printed at	:	**Replica Press**

Foreword

Human history has been defined by the production and use of energy. As human civilisation evolved our ancestors discovered fire which allowed them to cook food, preserve and consume it with enhanced benefits as well as to keep out the cold in their homes and thus be able to live with the elements. However, what brought about a major transformation of economic opportunities was the discovery of the steam engine and the ability of human society to generate power which prior to that invention was confined only to muscle power for performing specific production activities or for providing mobility. Since then rapid innovation and the progress of science and technology has made it possible for economic activities to reach unprecedented levels and add dimensions that were earlier unimaginable.

If we look at the future of energy we are likely to find a major transformation in the offing. This would be driven by concerns related to security of energy supply, the problem of pollution which occurs throughout the energy cycle and the growing challenge of climate change, which results from emissions and growing concentration of greenhouse gases including carbon di-oxide in the atmosphere, which is largely the outcome of growing dependence on fossil fuels. For these reasons a shift from fossil fuels and use of a totally different set of technologies involving energy are likely to replace current patterns and trends of energy supply and consumption. Another factor that would alter the energy scenario in the future is the current problem of lack of energy access for a large part of the human population, which is deprived of modern forms of energy. A total of 1.3 billion people in the world have no access to electricity and more than twice the number use biomass, often of very poor quality, as fuel for cooking and space heating.

Sustainable energy - energy that is accessible, cleaner and more efficient – powers opportunity. It grows economies; lights up homes, schools and hospitals; empowers women and local communities; and paves a path out of poverty to greater prosperity for all. Investing in sustainable energy is a smart strategy for growing markets, improving competitiveness, and providing greater equity and opportunity. Renewable resources are an important aspect of sustainable energy. The most frequently used renewable resources are solar, wind, biomass and geothermal. Besides, hydropower is one of the most widely used renewable energy resources in the world. The sun has been providing energy to Earth in the form of heat and light ever since the planet was formed. Wind is a renewable resource, usage of which is slowly growing globally and in the developing countries. Biomass resources include trees, food crops, algae, agricultural and forestry byproducts, and methane gas from landfills. Geothermal energy comes from harnessing heat from below the surface of the Earth. With the rising cost of fossil fuels due to depletion of reserves and increased demand for energy, renewable resources are receiving increasing attention.

The promise of renewable energy can only be realised through significant investments in research and development on alternative, sustainable technologies such as solar, biomass, wind, hydropower, geothermal power, ocean energy sources, solar-derived hydrogen fuel, and the energy storage technologies necessary to exploit all these competitively. Indeed, the transition to sustainable energy may well be the biggest business opportunity of the 21st century.

Improvements in energy efficiency are relatively quick and cost-effective ways to manage energy demand. Effective energy efficiency programmes can reduce a country's reliance on various energy sources used, and if these are imported in the country, they can in turn improve national security and stabilize energy prices. Smart grid options are emerging as important tools to manage energy efficiently. Improvements in energy efficiency represent the goal of efforts to reduce the amount of energy required to provide products and services.

I am delighted to learn that the Centre for Science and Technology of the Non-Aligned and Other Developing Countries (NAM S&T Centre) had organised the Third Triennial International Workshop on 'Sustainable Energy for All: Transforming Commitments to Action' in Bengaluru, Karnataka, India during 22-24 February, 2014 jointly with the Society of Energy Engineers and Managers (SEEM), Trivandrum, India, and in association with the Christ University Faculty of Engineering, Bangalore, India. The event was successful in highlighting the importance of Sustainable Energy by improving energy access, enhancing energy efficiency and increasing the share of renewable energy. The present book entitled **"Sustainable Energy for All: Transforming Commitments to Action - Lessons Learned and Actions for the Future"** is an outcome of these deliberations and comprises 15 scientific and technical papers from experts and professionals from 17 countries.

I compliment the NAM S&T Centre for bringing out this highly valuable publication. This is an important step forward that will help developing countries in working out appropriate strategies, policies and programmes for the exploitation of new technologies in the area of sustainable energy.

Dr. Rajendra K. Pachauri

Director General,
The Energy and Resources Institute (TERI),
New Delhi, India

Preface

The objective of this book is to present different dimensions of the Sustainable Energy as practiced in the participant countries to the workshop. The book is written to provide illustrations from countries on "Transforming Commitments to Action", which hopefully will provide valuable lessons for the future to all the countries. The book is expected to enable readers to reach a sound understanding of a broad range of topics related to the sustainable energy, including energy resource planning, renewable energy resources and programmes, and energy technology applications, including smart grid.

The book is recommended for engineers, researchers, energy planners and technical staff involved in energy efficiency programs.

Renewable resources are an important aspect of sustainable energy. The most frequently used renewable resources are biomass, water, wind, solar, and geothermal. Biomass resources include trees, food crops, algae, agricultural and forestry byproducts, and methane gas from landfills. Geothermal energy comes from harnessing heat from the Earth. A large utility can use a geothermal reservoir to run generators and produce electricity. The sun has produced energy in the form of heat and light since the Earth was formed. Solar energy systems do not produce CO_2 emissions and are often not harmful to the environment. Hydropower is one of the most widely used renewable energy resource in the world. Wind is a renewable resource available in a large number of countries. With the rising cost of fossil fuels due to increased demand for energy, renewable resources are receiving increasing attention.

Energy efficiency is a relatively quick and cost-effective way to manage energy demand and minimize depletion of resources. It is considered most important fuel to meet the energy requirements in near future. Effective energy efficiency programs can reduce a country's reliance on various energy sources used, and if these are imported in the country, it can in turn improve national security and stabilize energy prices. Smart grid is emerging as an important tool to manage energy efficiently. A smart grid is an electric grid system where all participants in the grid system (from electricity generators, to transmission and distribution operators, to electricity consumers) communicate and work with each other to increase the efficiency and reliability of the grid. A key feature of a smart grid system is the use of advanced technologies (such as smart meters) that provide participants with relevant, real-time information. The book provides clear, useful and attractive materials for the above mentioned topics.

The book is divided in three main parts; Energy Resources and Planning for Sustainability, Renewable Energy Programmes and Energy Efficient Technologies and Smart Grid. Each part contains papers related to the theme. Many experts from all over the world, from academicians to planners and implementers contributed to this book. The aim is to provide knowledge, methods and best practices and lessons learnt in the area of sustainable energy.

Section I: Energy Resources and Planning for Sustainability

This part includes papers on general theme related to sustainable energy, encompassing the whole energy sector in the countries. The chapter 1 on sustainable energy for all brings out the context of the initiative and broader action agenda at global level. Chapter 2 on Indonesia covers energy sector in detail with policy and regulatory framework and renewable energy development in Indonesia. The Mauritius study in chapter 3 brings out sustainable energy planning at the national level. Finally, affect of irrigation, energy and environmental factors on agricultural productivity in Nigeria is included in chapter 4, last chapter of this part.

Section II: Renewable Energy Programmes

Renewable energy has been an important cornerstone of sustainable energy and renewable energy programmes have been implemented in several countries. This part brings out salient features of renewable energy resources and programmes in the countries. A comprehensive assessment of bio-fuel resources and technologies has been covered in the Chapter 5 on India. A portable biogas power plant as pilot in Iraq is covered in the Chapter 6. This is followed by the study of renewable energy usage in the Dost Mimaropa region in Philippines in Chapter 7. It is followed by current status of renewable energy in Vietnam and approaches to address the issues in its development in Chapter 8. Research on renewable energy in Zambia is covered in the Chapter 9. Pattern of usage of renewable energy in institutions is included in the Chapter 10. A case study of photovoltaic project in Zimbabwe is discussed in chapter 11 with critical comments on stakeholder involvement process and technology transfer.

Section III: Energy Efficient Technologies and Smart Grid

This part covers information about advanced energy technologies and their applications through case studies in the context of sustainable energy. First chapter in this part (chapter 12) describes the monitoring and novel applications for smart grid system installed on real grid. This is followed by interesting case study of energy assessment and implementation of cleaner production in a tea factory in Sri Lanka in chapter 13. The next chapter describes high efficiency mix energy system design with low carbon footprint in which a furnace saves energy by using oxygen-enriched combustion in Thailand, a novel idea that can be replicated elsewhere. Last chapter (chapter 15) brings out efficient use of energy in district heating system of Soma Province in Turkey, through utilization of waste heat from power plant.

Moustafa Mohammed Eissa

R. Hari Kumar

Jyoti Prasad Painuly

Introduction

Sustainable energy - energy that is accessible, cleaner and more efficient – powers opportunity. It grows economies; lights up homes, schools and hospitals; empowers women and local communities; and paves a path out of poverty to greater prosperity for all. In short, it is meeting the needs of the present without compromising the needs of the future.

The world currently invests more than $1 trillion per year in energy, much of it going toward the energy systems of the past instead of building the clean energy economies of the future. Investing in sustainable energy is a smart strategy for growing markets, improving competitiveness, and providing greater equity and opportunity. The promise of renewable energy can only be realised through significant investments in research and development on alternative, sustainable technologies such as solar, biomass, wind, hydropower, geothermal power, ocean energy sources, solar-derived hydrogen fuel, and the energy storage technologies necessary to operate them competitively. Indeed, the transition to sustainable energy may well be the biggest business opportunity of the 21st Century.

'Sustainable Energy for All (SE4All)' is a global initiative launched by the Secretary-General of the United Nations, Ban Ki-moon, in the year 2011. It is aimed at achieving energy access, improving energy efficiency and increasing the use of renewable energy. Under this initiative it is intended to double the rate of improvement in energy efficiency, double the share of renewable energy in the global energy mix, and achieve universal energy access by the year 2030.

In order to deliberate on variety of issues concerning the sustainable energy, the Centre for Science and Technology of the Non-Aligned and Other Developing Countries (NAM S&T Centre) jointly with the Society of Energy Engineers and Managers (SEEM), Trivandrum, India, and in association with the Christ University Faculty of Engineering, Bangalore, India organised the Third Triennial International Workshop on 'Sustainable Energy for All: Transforming Commitments to Action' in

Bengaluru, Karnataka, India during 22-24 February 2014. Mangalore Refinery and Petrochemicals Limited, a subsidiary of the Indian Oil and Natural Gas Corporation Ltd. (ONGC), was the Principal Sponsor of this workshop. The State Designated Agencies (SDA) of the Bureau of Energy Efficiency (BEE) in Karnataka, the Karnataka Renewable Energy Development Limited (KREDL) and the Government of Karnataka undertaking Power Company of Karnataka Limited (PCKL) co-sponsored this event.

70 experts from 19 countries, namely, Australia, Denmark, Egypt, Indonesia, Iraq, Malaysia, Mauritius, Myanmar, Nigeria, The Philippines, Sri Lanka, Thailand, Turkey, USA, Vietnam, Zambia, Zimbabwe, and the host country India participated in this International Conference. 26 scientific papers and country status reports were presented during the Conference. The International Renewable Energy Agency (IRENA) with its office in Abu Dhabi, UAE was represented as an Observer.

The present book is an edited compilation of 15 papers read during the Conference and found suitable for this publication. At the end of the book is placed a copy of the 'Bengaluru Declaration on Sustainable Energy for All – 2014' that was unanimously adopted by the Conference participants.

I gratefully acknowledge the dynamic involvement and untiring effort of Dr. Jyoti Prasad Painuly, Head, SE4ALL Energy Efficiency Hub, UNEP Risø Centre, Technical University of Denmark; Prof. Moustafa Mohammed Eissa of Helwan University, Cairo, Egypt; and Dr. R. Harikumar, Head - Education and Training Division Energy Management Centre, Govt of Kerala, India for technical editing of this publication. Last but not the least, valuable services provided by the entire team of the NAM S&T Centre, in particular, Mr. M. Bandyopadhyay for overall supervision, and Ms. Shania Tahir, Ms. Subhasree Basu and Mr. Pankaj Buttan in compiling the papers, coordinating between the editors and the authors and giving shape to this volume are greatly appreciated.

I hope that the information provided and suggestions and recommendations made in the papers included in this book will be a good reference material on Sustainable Energy for all.

Prof. Dr. Arun P. Kulshreshtha
Director General,
NAM S&T Centre

Contents

Section I
Energy Resources and Planning for Sustainability

Section II
Renewable Energy Programmes

Section III
Energy Efficient Technologies and Smart Grid

SECTION I
Energy Resources and Planning for Sustainability

Chapter 1

Agenda and Action Programme: Sustainable Energy for All Initiative

Jyoti Prasad Painuly

Copenhagen Centre on Energy Efficiency, UN City,
Copenhagen, Denmark
E-mail: jypa@dtu.dk

ABSTRACT

The UN Secretary General announced the SE4ALL initiative in 2011, with three interlinked objectives: to provide access to modern energy to all by 2030, double the share of renewable in global energy mix, and double the improvement rate of energy efficiency by 2030. UN and The institutional structure to guide the process has evolved with identification of agencies responsible for actions in countries, as well initiation of efforts to make enabling environment for private sector participation, whose engagement is crucial to raise the necessary finances and to achieve the targets. The institutional structure includes regional hubs for country actions and thematic hubs to provide crucial support to the initiative. Recent developments indicate that process has started on the right note with stakeholders from governments, private sector, UN system, civil society and others quite enthusiastic and supportive of the initiative. It will however be important to increase the momentum and engage more and more stakeholders, primarily governments and private sector in the initiative. Achievement of the SE4ALL target may help contain the global temperature to 2 degree Celsius, a much sought after goal to prevent serious climate change. The paper provides an overview of the SE4ALL initiative, including recent developments, with particular focus on energy efficiency.

Keywords: *Sustainable energy, Energy efficiency, SE4ALL, Energy efficiency hub, Private sector, Accelerators.*

1. INTRODUCTION

The World Energy Outlook (WEO, 2012) highlights the issues related to sustainable development with the observation that "taking all new developments and policies into account, the world is still failing to put the global energy system onto a more sustainable path". Despite progress made, nearly 1.3 billion people remain without access to electricity and 2.6 billion do not have access to clean cooking facilities. The WEO also emphasises the fact that energy efficiency is recognized as a key option in the hands of policy makers but current efforts fall well short of tapping into its full economic potential. Governments are becoming increasingly aware of the multiple benefits of energy efficiency improvements in achieving sustainable growth and energy efficiency is now a primal area in the climate change mitigation agenda, with the potential to reduce global emissions by 1.5 Gt CO_2 in 2020. The renewables, on the other hand, with steady increase in hydro, wind and solar energy are important contributor to the reduction in fossil fuel emissions and helping the world move towards sustainable energy.

Contributing to this positive momentum, the UN Secretary-General Ban Ki-Moon launched, in 2011, the Sustainable Energy for All (SE4ALL) global initiative to mobilize action from all sectors of society in support of three interlinked objectives to be achieved by 2030:

☆ Ensure universal access to modern energy services;

☆ Double the global rate of improvement in energy efficiency; and

☆ Double the share of renewable energy in the global energy mix

Achieving the three objectives together will maximize development benefits and help stabilize climate change over the long run.

The initiative seeks to build a global multi-stakeholder partnership between governments, the private sector, and civil society to achieve the three objectives. The goals, actions, projects and programmes of SE4ALL are supported by various stakeholders including development banks, international organisations, private sector and various civil society organizations.

Global investment in the areas covered by the three SE4ALL objectives has been estimated at $400 billion in 2010. The investments required to achieve the three objectives are tentatively estimated to be at least $600–800 billion per year over and above existing levels, entailing a doubling or tripling of financial flows over current levels. According to some experts, the bulk of those investments are required to deliver the energy efficiency and renewable energy targets.

Efforts on energy access are being led by several agencies including World Bank, EU and country governments in Asia and Africa. Renewable has also been a focus area for several years, resulting in healthy growth of renewable energy at the global level, with $318 billion investment in renewables in 2011, though with some decline in 2012 and 2013. Energy efficiency has remained neglected considering its potential, and focus of this paper is on energy efficiency objective of the SE4ALL. Implementing all currently available measures with reasonable payback periods would be enough to meet or even exceed the SE4ALL objective. However, there are currently several barriers hindering the broad adoption of energy efficiency initiatives.

According to the SE4ALL Global Tracking Framework (WB, 2012) study lead by the World Bank and the IEA with the support of various other institutions, energy intensity decreased at a compound annual growth rate (CAGR) of 1.3 percent over the 20 years between 1990 and 2010, as a result of increased energy efficiency. However, the rate of improvement in energy efficiency slowed considerably during the period 2000–2010, showing a CAGR of 1.0 percent, compared to 1.6 percent per year for 1990–2000. With the starting point for measuring future progress in global energy efficiency under the SE4ALL set as a CAGR for global energy intensity of 1.3 percent, the SE4ALL global objective is therefore a CAGR of 2.6 percent for the period 2010–2030. Energy intensity has been used as proxy for energy efficiency due to constraints in measuring and estimating energy efficiency at macro levels.

Doubling the rate of improvement in energy efficiency is a formidable challenge, as it requires significant mobilization of resources through public-private partnerships (PPPs), and needs to overcome a variety of barriers that impede progress in energy efficiency. Some of these barriers contributing to the lack of uptake of energy efficiency relate to policy, regulatory, financial, technical, institutional, capacity, and subsidies.

The world can benefit greatly from the initiative, if the targets are achieved. According to a study carried by IIASA (2013), achievement of targets can help contain the global temperature rise to 2 degrees Celsius, a target widely espoused by climate scientists, IPCC and others to prevent serious damage from climate change.

2. SE4ALL INSTITUTIONAL FRAMEWORK

SE4ALL Initiative Organisational Structure

At the apex level, the initiative has an Advisory Committee. The committee is co-chaired by the UN Secretary-General and the President of the World Bank and consists of representatives from UN organisations, private sector and civil society to provide guidance for the SE4ALL operations. An Executive Committee provides operational oversight and is headed by the Chairman of the Bank of America. The Secretary-General's Special Representative for SE4ALL is the chief executive officer (CEO) of the initiative and is supported by a Global Facilitation Team (GFT). The Advisory Committee and the Executive Committee are supported by four thematic committees – Energy Access Committee, Energy Access Committee, Renewable Energy Committee, and Finance Committee (a cross-cutting theme).

The SE4ALL Organisational Structure is shown in Figure 1.1.

Each of the three committees related to SE4ALL targets provide high-level support to the respective target; for example, the energy efficiency committee provides support to the target of doubling the global rate of improvement in energy efficiency by 2030. The finance committee's mandate is to help raise the necessary finances across the targets. The committees operates under the overall guidance and direction of the advisory board of SE4ALL, and complement each other's activities in efforts to achieve the SE4ALL targets. The GFT, based in Vienna, is headed by the UN Secretary General's

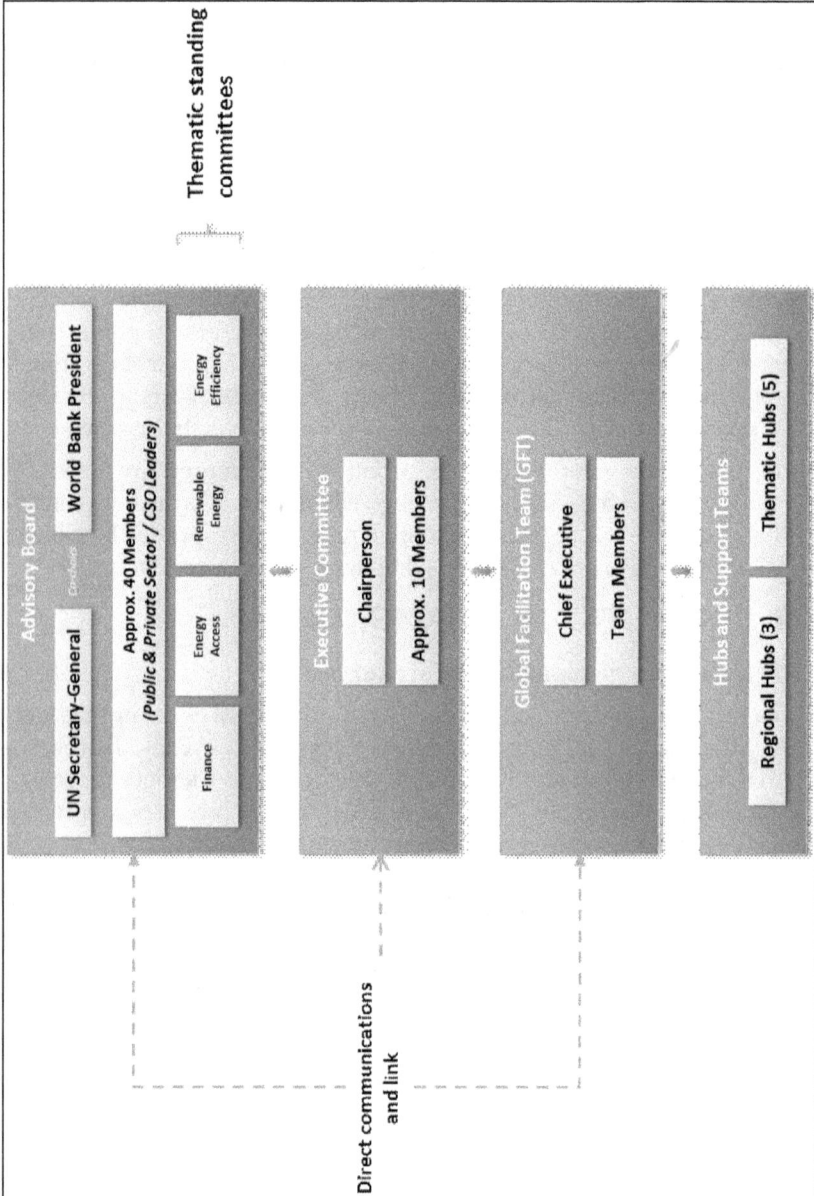

Figure 1.1: Sustainable Energy for All (SE4ALL) Initiative Organizational Structure[1]

1 The SE4ALL Organisational Structure presented refers to the organisational structure in place as of 30 May 2014

special representative, who is also CEO of the SE4ALL initiative. The GFT acts as the secretariat of the SE4ALL initiatives and facilitates actions of the various stakeholders, including Regional Hubs, Thematic Hubs, UN organisations, civil society and others. The GFT has four focal points to facilitate the work of the committees and the hubs in the four focus areas of the SE4ALL[2].

Country Actions and Support Structure

Actions in countries to achieve SE4ALL targets are envisaged to be led by the World Bank (across the regions) and by three regional development banks – African Development Bank (AfDB), Asian Development Bank (ADB) and the Inter-American Development Bank (IADB) – in their regions alongside the World Bank.

Within the World Bank, country actions are being led by the Energy Sector Management Assistance Programme (ESMAP), who carried out rapid assessments in more than 40 countries to map the status and identify gaps in achieving the three SE4ALL objectives. ESMAP is currently working with several countries with a focus on energy access and mobilizing investment.

A list of 30 priority countries has been drawn up by the Global Facilitation Team for early action; the list is available at the SE4ALL website.

Thematic hubs have been envisaged to provide necessary support for SE4ALL targets globally as well as in the countries requiring such support. Concept of thematic hubs is described in the section that follows.

Concept of Hubs under the SE4ALL Initiative

The SE4ALL Strategic Work Program introduces the concept of 'hubs' as critical mechanisms that help focus efforts within the institutional structure of the SE4ALL initiative. A hub acts as a platform that facilitates action and engagement in a specific region or thematic area critical to the achievement of the SE4ALL targets.

The Regional hubs are conceptualised to provide support to countries in their activities and actions to achieve SE4ALL targets.

Thematic Hubs are conceptualised to provide crucial support for the three objectives of the SE4ALL. The three thematic hubs initially conceptualised included a Knowledge Management Hub to provide support on issues requiring development of new knowledge for SE4ALL work, a Renewable Energy Hub to provide thematic support to renewable energy target, and an Energy Efficiency Hub to provide thematic support to energy efficiency target.

The renewable energy hub is housed by the International Renewable Energy Agency (IRENA), and the energy efficiency hub is hosted by the Copenhagen Centre on Energy Efficiency (C2E2), both the hubs focusing on information sharing, forging partnerships, and providing a platform for cooperation around specific action areas on renewable energy and energy efficiency respectively. With the financial support from Danish Government, the energy efficiency hub was established by the Danish

2 Energy Access, Energy Efficiency, Renewable Energy, and Finance.

Government, UNEP and Denmark Technical University in the C2E2, which is a part of the university. The knowledge management hub hosted by the World Bank focuses on facilitating dialog on global energy knowledge and multi-stakeholder partnerships needed to meet the SE4ALL objectives. The hub will also collaborate with thematic and regional Hubs in areas of intersection and provide support through the development of cross-cutting knowledge products. Subsequently, more hubs have been announced that include a Bottom-Up Energy Solutions Hub (hosted by UNDP) and a Capacity Building Hub (hosted by TERI). The capacity building hub was launched in June 2014 and the bottom-up energy solutions hub is also expected to be launched in 2014.

3. ENERGY EFFICIENCY HUB WORK PROGRAMME

The energy efficiency hub is envisaged to provide support to achieve energy efficiency objective of the SE4ALL programme. The initial hub programme, described below, has been prepared after consultation with various stakeholders and is tailored to provide this support. Primary tasks as a part of this support include:

Championship on Energy Efficiency

A need for championing cause of energy efficiency globally was identified. It would require raising awareness on energy efficiency and advocacy of the same at various forums. The hub will be providing necessary support to the UNEP for energy efficiency championship, for which UNEP is positioned well with its global outreach and a few ongoing global initiatives on energy efficiency in partnership with various agencies. Private sector engagement in the SE4ALL initiative is crucial, and among others, UNEP has also been involved in helping bring the private sector on board. The hub has been providing necessary support to UNEP for this.

As a part of the task, the hub will also analyse and highlight pathways to achieve the energy efficiency target of the initiative, so that the best course of action becomes visible. The information can be used by countries to realign action and set priorities.

Analytical and Technical Support

The hub will provide analytical support to the initiative through a variety of analyses, which may include items like assessment of national, regional and global strategies, commitments and actions plan in the area of energy efficiency, analysis of high impact opportunities and contribution they can make at country and global level, and so on.

The hub will also provide technical support to the developing countries in collaboration with regional hubs and other agencies on issues related to energy efficiency: for example estimating energy efficiency potential, preparing action plans, identifying and addressing barriers etc.

Knowledge Platform on Energy Efficiency

The hub is developing a knowledge platform on energy efficiency where stakeholders can get access to information on energy efficiency, including on databases and experts, forge partnerships and collaborate, share experiences including best practices.

4. CURRENT DEVELOPMENTS

Since the launch of the initiative in 2011, initial efforts were focussed on developing the institutional structure, putting required frameworks in place for tracking and monitoring the progress, mobilizing investments, and country actions. Major developments include the following:

Global Tracking Framework (GTF)

The GTF has been developed by the ESMAP and IEA in collaboration with 13 other agencies to track and monitor progress at the country levels in three action areas of the SE4ALL- energy access, energy efficiency and renewable. It uses indicators to track the progress. The data from the countries will be consolidated at the global level and analysed to assess the reach of energy access, the share of renewables, and the rate of improvement of energy efficiency to check whether the world is on track to achieve the objectives of Sustainable Energy for all. The framework is currently being updated as a part of biennial updation exercise.

Readiness for Investment in Sustainable Energy (RISE)

RISE framework helps prepare a prospectus in countries to remove barriers and promote investments to sustainable energy. The SE4ALL targets achievements require trillions of dollars in global investment. Private sector investment is a must alongside public investment to reach the investment levels required to achieve the targets. The RISE indicators measure the extent of enabling environment for the investment in a country, thereby helping policy makers to take suitable action. The framework has been developed by the World Bank in collaboration with other partners.

Country Actions

ESMAP carried out rapid assessment in more than 40 countries to identify gaps in the three action areas of SE4ALL with a view to help countries prepare action plans accordingly. More than 80 developing countries have joined the SE4ALL initiative and GFT identified 30 priority countries for initial country actions. Template for country action plans have been finalised and two regional workshops have been held by the AfDB in Africa to help countries initiate preparation of country action plans. ESMAP is working in 10 countries, primarily on energy access. IADB has also initiated action in Latin America. ADB launched the Asia Regional Hub in the recently concluded clean energy forum in June 2014 and expected to start action in countries soon. Among other stakeholders, EU has launched country action programmes in several countries with a funding of Euro 50 million with their own framework to support the initiative.

Progress on Energy Efficiency

In March 2014, a meeting was held in Paris, convened by the CEO Accenture, co-chair of the energy efficiency committee, to explore course of action to accelerate energy efficiency improvement globally. Concept of accelerator was developed in the meeting, which consisted of developing a private-public partnership platform around a few identified high impact opportunities. Referred as "Global Energy Efficiency Accelerator Platform", its initial focus is on lighting and appliance, district heating, building, and transport sectors, for which "accelerators" have been developed. Industrial energy efficiency has also been added to the list recently. A brief description of the concept of the platform is given in Box 1.

Box 1: Global Energy Efficiency Accelerator Platform

The Global Energy Efficiency Accelerator Platform has been established to help reach the SE4ALL energy efficiency objective. It will do so by driving action and commitments by national and sub-national leaders at the country, city, state, region, or sector level. A key deliverable will be Integrated Policy and Investment Roadmaps prepared with committed public and private partners. These Roadmaps will guide project implementation supported by a global network of experts, institutions and businesses.

Source: SE4ALL Energy Efficiency Accelerator Platform Flyer

The accelerator approach including focus on four to five accelerators in the initial stage was endorsed by the energy efficiency committee in a meeting hosted by the hub on May 12, 2014 in Copenhagen. A meeting of Danish industries was organised by the hub in early March with the view to solicit opinion of the private sector and engage them in various initiatives on energy efficiency. An energy efficiency stakeholders' workshop entitled "Doubling the Global Rate of Improvement in Energy Efficiency by 2030; Options, Implementation Issues and Way Forward" was also organised by the hub on June 16-17, 2014. The workshop was attended by a cross-section of experts, government officials, UN agencies, private sector, and civil society, and representatives included both from developing and developed countries from various regions.

Other Progress

A Knowledge Exchange Forum as organised by the ESMAP in October 2013 to share experiences on SE4ALL work done by various stakeholders. IRENA has developed the model REMAP and worked out level of effort and investment required to double the share of renewables in 26 countries. The SE4ALL Global Forum was held from June 4 to 6, 2014 in New York to "assess progress on sustainable energy since Rio, showcase success, share best practices, present new commitments, catalyse action and help shape the global energy debate for the next decade"[3]. The opportunity was thus used to take a stock of the situation, discuss way forward, make partnerships and renew commitments to SE4ALL goals by the world leaders.

5. CONCLUSIONS

The SE4ALL is a unique initiative in which UN has joined hand with private sector and civil society to address three crucial issues with targets for 2030; to provide access to modern energy to all by 2030, double the share of renewable in global

---

3 http: www.se4all.org/se4all-forum/

energy mix, and double the improvement rate of energy efficiency. The institutional structure to guide the process has evolved with identification of agencies responsible for actions in countries, as well initiation of efforts to make enabling environment for private sector participation, whose engagement is crucial to raise the necessary finances and to achieve the targets. The institutional structure includes regional hubs for country actions and thematic hubs to provide crucial support to the initiative. Achievement of the SE4ALL target may help contain the global temperature to 2 degree Celsius, a much sought after goal to prevent serious climate change. Recent developments indicate that process has started on the right note with stakeholders from governments, private sector, UN system, civil society and others quite enthusiastic and supportive of the initiative. It will however be important to increase the momentum and engage more and more stakeholders, primarily governments and private sector in the initiative.

Abbreviations

AU:	African Union
ADB:	Asian Development Bank (ADB)
AfDB:	African Development Bank (AfDB)
CAGR:	Compound annual growth rate
C2E2:	Copenhagen Centre on Energy Efficiency
CEO:	Chief Executive Officer
ESMAP:	Energy Sector Management Assistance Programme
EU:	European Union
GFT:	Global Facilitation Team
GTF:	Global Tracking Framework
IADB:	Inter-American Development Bank
IEA:	International Energy Agency
IIASA:	International Institute of Applied System Analysis
IRENA:	International Renewable Energy Agency
RISE:	Readiness for Investment in Sustainable Energy
SE4ALL:	Sustainable Energy for ALL
TERI:	The Energy and Resources Institute
UN:	United Nations
UNDP:	United Nations Development Programme
UNEP:	United Nations Environment Programme
WB:	World Bank
WEO:	World Energy Outlook

REFERENCES

1. IEA, 2013. Redrawing the Energy-Climate Map, International Energy Agency.

2. IIASA, 2013. UN Sustainable Energy Initiative Could Help Limit Climate Change,

3. http: www.iiasa.ac.at/web/home/about/news/NCC-UN-SE4All.en.html

4. SE4ALL Documents: from the website: http: www.se4all.org/

5. SE4ALL. Accountability Framework, Sustainable Energy for ALL. http: www.se4all.org/tracking-progress/accountability-framework/

6. SE4ALL. Technical Documents, Sustainable Energy for All. http: www.se4all.org/resources/technical-documents/

7. SE4ALL, 2012. Sustainable Energy for All- A Framework for Action The Secretary-General's High-level Group on Sustainable Energy for All

8. SE4ALL. SE4ALL Global Tracking Framework, http: www.se4all.org/2013/10/29/se4all-global-tracking-framework/

9. RISE, 2013. Presentation by the WB team in the RISE Advisory Group Meeting of the Energy Efficiency.

Chapter 2

Renewable Energy and Energy Efficiency Development in Indonesia

Soni Solistia Wirawan, Oo Abdul Rosyid
and Hari Yurismono

Energy Technology Center - Agency for the Assessment and
Application of Technology (B2TE-BPPT), Indonesia
E-mail: soni.solistia@bppt.go.id

ABSTRACT

In the last 10 years, Indonesian economy grew at an average 6 per cent per year and the population increased by about 1.49 per cent per year. It resulted in an increased average energy consumption of 6.2 per cent per year. Currently, the energy consumption is still dominated by fossil energy (oil, gas, and coal), which has limited resources and harms the environment. Meanwhile, development of new and renewable energy (NRE) and energy efficiency in Indonesia is l relatively slow, despite large potential of renewable energy and importance of its role in the national energy mix. At present, the contribution of renewable energy in the national energy mix is around 5 per cent. Indonesian government seeks to increase the use of renewable energy to 25 per cent by 2025, and reduce greenhouse gas (GHG) emissions by 26 per cent by 2020. To encourage the development and utilization of renewable energy, and energy efficiency, the GOI government has issued a number of policies and regulations, including regulation No. 5/2006 on National Energy Policy. But there are still some problems in implementation. BPPT as one of the Indonesian institutions, that focussed on national energy issues, is addressing the issue by developing and encouraging the implementation of renewable energy and energy conservation technology. Therefore, many efforts are required for dissemination and harmonization of the existing national policies, along with strategies for the achievement of renewable energy and energy efficiency development in Indonesia. This paper discusses the current state and development of renewable energy and energy efficiency development in Indonesia.

Keywords: *Indonesia, Renewable energy, Fossil fuel, Environment impact, Energy mix, Energy efficiency, Energy policy.*

1. INTRODUCTION

Indonesia is an archipelago nation located in South East Asia. It is the fourth most populous country in the world and is home to the world's largest Muslim population. Its 13,466 islands foster a diverse range of cultures, languages and ethnicities, which inspired the country's motto of "unity in diversity". Indonesia was once a Dutch colony and declared its independence shortly after Japanese occupation ended in 1945. After a long period of political instability, Indonesia held legislative elections in 1999 and is now regarded as the third-largest democracy in the world. Important issues within Indonesia include combating poverty, the implementation of reforms and the promotion of transparency in its governmental institutions.

Indonesia as a developing country with energy growth of 7 per cent per year, and have not been offset by growth in energy supply. While the population of Indonesia, which now reaches up to 244.5 million people. All of this growth is surely accompanied by the increase in energy demand due to the increasing number of homes, factories, and commercial and industrial buildings. If we assume that demand for electricity will grow in average 7 per cent per year for the next 30 years, then electricity consumption will significantly increase, for example in the household sector, consumption will increase from 21.52 GWh in 2000 to around 444.53 GWh in 2030. There are four main sectors of energy users, namely household, commercial, and industrial and transportation sector. Currently, the largest energy user is the industrial sector with a share of 44.2 per cent. Next largest consumption is the transportation sector with 40.6 per cent, followed by the household sector with 11.4 per cent and the commercial sector with 3.7 per cent (Energy Outlook, 2013).

Until now, the primary sources of energy still come from fossil fuels, with 46.9 per cent from oil, 26.4 per cent from coal, and 21.9 per cent from natural gas. Hydro (water) power and other renewable energy only make up about 4.8 per cent from the total of utilized energy resources. In fact, fossil energy reserves are being depleted. Various environmental problems as a result of the byproducts of combustion of fossil fuels can cause greenhouse effect and global warming. Along with the increase in the rate of energy consumption and the depleting fossil energy production capability in the country requires us to take steps in using energy efficiently, finding inexpensive alternative energy sources, sustainable, and environmentally friendly. Application of renewable energy as an alternative energy that can provide energy services and products that are reliable, sustainable, clean, and competitive to help meet growing global energy needs growing rapidly. Indonesia has the potential of new and renewable energy is quite large and varied. However, utilization is still not optimal, and contribute to the national energy mix is less than 5 per cent. The Indonesian government seeks to increase the use of renewable energy to 25 per cent by 2025, and reduce greenhouse gas (GHG) emissions by 26 per cent by 2020. To encourage the use of new and renewable energy and energy efficiency in Indonesia, the government has issued Presidential Regulation No.5/2006 on national energy policy, and other legislation that supports the development of new and renewable energy. This paper discusses the current state of energy, regulations, development of renewable energy and energy efficiency in Indonesia.

2. MATERIALS AND METHODS

The data source of this study was obtained from several references mainly from the Ministry of Energy and Mineral Resources in the form of laws, government regulations, ministerial rules relating to new and renewable energy policy, and various BPPT's energy studies (Energy Outlook, Renewable Energy and Energy Efficiency Studies). While the data related to the development of new and renewable energy technologies obtained from other related resources. The data or information is then analyzed to obtain information about the current status of renewable energy and energy efficiency development in Indonesia, constraints and prospects, and the extent to which support government policies and strategies in encouraging the development of renewable energy in achieving the target of 25 per cent in the national energy mix by 2025.

3. RESULTS AND DISCUSSION

3.1 Energy Supply

Energy is an essential part of everyday life. Virtually all of our everyday activities and productive processes involve energy in one form or another. Fuels and electricity power our transport systems, heat our buildings and produce the goods and services. Because energy is such an essential part of almost all our activities, a secure, sustainable and affordable supply is vital to ensuring Indonesia's continued economic and social well-being. We meet our energy needs through a wide range of sources, both renewable (hydro, geothermal, wood, wind, biogas and solar) and non-renewable (fossil fuels such as oil, gas and coal). Around two-thirds of our energy supply is non-renewable, and a growing proportion of our electricity is generated from non-renewable sources.

3.1.1 Primary Energy Supply

Indonesia is endowed with large hydrocarbon and renewable energy resources. Indigenous oil, gas and coal reserves have played an important role in Indonesia's economy as a source of energy, industrial raw material and foreign exchange earner. Since 2000, domestic consumption has continued to rise while production has been falling, so in recent years Indonesia has begun imported, especially oil.

Oil resources in Indonesia are estimated to be 56.6 billion barrels, with proven reserves of about 8.4 billion barrels (Table 2.1). With the oil production level in the year 2007 approximately 348 million barrels per year, it is estimated that the reserves will be exhausted within the next 24 years, if there are no new reserves discovered. Natural gas resource at the same time is approximately 334.5 TCF with a proven reserve of about 165 TCF. With a production rate of approximately 2.79 TCF in 2007, it is estimated that gas reserves will be exhausted within the next 59 years. While coal reserves are estimated at about 90.5 billion tons, the proven reserves estimate is of 18.7 billion tons. If the production rate was constant at 201 million tons as in the year 2007, it is estimated that the reserves will be exhausted in 93 years.

Table 2.1: Resources and Reserves of Fossil Energy in Indonesia

Fossil Energy	Resources	Reserve	Production	RP Ratio (year)
Oil	56.6 billion barrel	8.4 billion barrel*	348 million barrel	24
Gas	334.5 TSCF	165 TSCF	2.79 TSCF	59
Coal	90.5 billion ton	18.7 billion ton	201 billion ton	93
Coal Bed Methane	453 TSCF	–	–	–

Source: MEMR, 2010

Oil defines a major sector in the Indonesian economy. During the 1980s, Indonesia was a significant oil-exporting country. Since 2000, domestic consumption has continued to rise while production has been falling, so in recent years Indonesia has begun importing increasing amounts of oil. Within Indonesia, there are considerable amounts of oil in Sumatra, Borneo, Java, and West Papua Province. There are said to be around 60 basins across the country, only 22 of which have been explored and exploited. Prior to 2002, Indonesia had been a net energy exporter of oil, gas and coal. Due to increased demand, depleted reserves and failure to develop new, comparable resources, Indonesia has commenced the import of crude oil and refinery products and has been a net importer of oil since 2004. It was a member of the Organization for Petroleum Exporting Countries (OPEC) from 1962 to 2009, but suspended its OPEC membership in January 2009.

Table 2.2: New and Renewable Energy Potentials in Indonesia

Non Fossil Energy	Resources	Installed Capacity
Hydro power	75.67 GW	4.2 GW
Geothermal	27 GW	0.992 GW
Mini/micro-hydropower	0.45 GW	0.084 GW
Biomass	49.81 GW	0.31 GW
Solar energy	4.8 kWh/m²/day	0.008 GW
Wind energy	9.29 GW	0.0005 GW
Uranium	3 GW (e.q. 24.112 ton) for 11 years	30 MW

Source: MEMR, 2010

Unlike fossil energy sources that have been widely used, renewable energy resources such as hydropower, geothermal, biomass, solar, and wind energy, have not been intensively utilized. The hydropower potential of 75.67 GW has been utilized at about 4.2 GW or around 6 per cent of the total potential (Table 2.2). Geothermal potential of 27 GW is currently utilized for only about 1.2 GW. The remaining potential of both hydropower and geothermal is relatively large. Other renewable resources such as, biomass, solar and wind energy potential is still abundant and not yet utilized optimally. In addition to above energy resources, it is estimated that a quarter of Indonesia's land contains radioactive mineral deposits, especially uranium which could be used as fuel for nuclear power plants in future.

Renewable energy sources have great potential for use in electricity generation. But until now utilization of renewable energy is very small, because there are some constraints in terms of technology and financing. Therefore, there is urgent need to push the government policy in order to make the renewable resources compete with fossil fuels for energy generation.

3.1.2 Electricity Supply

Electricity supply to the national electricity grid comes from generating capacities of the state owned, state electric company (PLN), IPPs (independent power producers), and captive power through power purchases. PLN is the single - buyer of IPP power and captive power purchase. In 2009, Indonesia's total power generating capacity to the grid was 31,453 MW. Coal power generation produced 41.8 per cent of total grid power in 2009. The shares of primary energy utilized by PLN, IPP power generation and captive power purchase are as shown in Figure 2.1. In 2009, total electricity generated was 155.3 TWh, of which 77.7 per cent was produced by PLN and total electricity sold was 134.5 TWh. In 2009, electricity use was mostly in the residential sector 40.8 per cent; followed by industry 34.3 per cent; commercial sector 18.4 per cent; and Other 6.4 per cent. Electricity supply to the commercial sector increased most rapidly at an average annual growth rate of 10.0 per cent; followed by residential sector, 6.8 per cent; and the industry sector 3.5 per cent. Indonesia however has substantial self generation or captive power in industry.

The PLN, owns most of the generation assets, transmission and distribution infrastructure. Two wholly owned subsidiaries of PLN, *Pembangkit Jawa-Bali* (PJB) and *Indonesian Power* (IP), are now the primary generators of electricity. Overall, PLN effectively controls 86 per cent of generation assets. Independent Power Producers

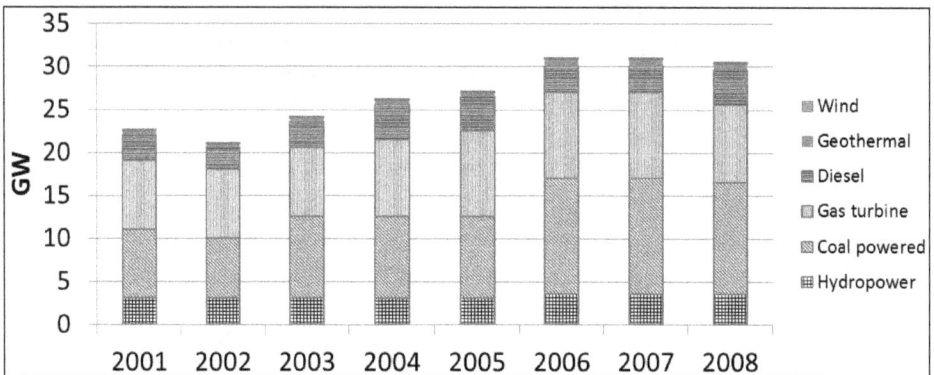

Figure 2.1: Shares of Primary Energy Utilized by PLN, IPP and Captive Power (MEMR, 2010)

(IPPs) and various electricity cooperatives also generate electricity, however the PLN is the sole buyer of electricity and retains the right of first refusal. The low rate of electrification in Indonesia is largely due to distribution and transmission infrastructure shortages. Experts have warned that unless Indonesia's electricity infrastructure is updated, then the country could face an energy crisis within a decade. Five distribution entities now act semi-autonomously from PLN to distribute electricity to the east, west, central Java, Bali and Jakarta regions. The Indonesian interconnected system encompasses Java, Bali and Madura. The islands of Java and Sumatra are connected by a 39 km submarine cable. Despite the increased privatization of the electricity sector, PLN controls all transmission infrastructure.

In 2006, the PLN served 54 per cent of Indonesia's households. PLN expects that 93 per cent of households will be connected to grid electricity by 2020. The Ministry of Energy and Mineral Resources reported that Indonesia's electrification ratio had increased to about 72 per cent in 2012. The electricity sector is expected to expand coal-fired generation, and to eliminate most use of oil. The share of electricity generated by oil will reduce from 31 per cent in 2005 to 4 per cent in 2015, and to 1 per cent by 2030. Electricity generated by coal-fired power plants is expected to increase to 64 per cent of all electricity generated in 2015; the final figure for 2030 is 67 per cent.

About 80 per cent of the electricity is consumed on Java and Bali alone. In recent years consumption of electricity has increased by 7 per cent annually. It is calculated that for every 1 percent increase in GDP the energy demand increases by 1.8 percent. Indonesia failed to meet this demand growth with adequate system investments which has resulted in increased frequency and duration of power outages which prove costly to local industries. These factors have sharply put the need for diversification of supplies into focus and Indonesia has an ambitious plan for renewable energy and in parallel are advancing plans for the use of nuclear energy.

3.2 Energy Demand

Final energy consumption including biomass was 946.0 million BOE in 2009, consumption increased at an average annual rate of 2.3 per cent over the past decade from 2000, the trend of final energy consumption is as shown in Figure 2.2. Residential sector final energy consumption was 314.7 million BOE in 2009, increasing at an average annual rate of 0.7 per cent from 2000 - 2009. Residential energy consumption saw structural change over the period specifically, kerosene consumption declined from 63.2 million BOE to million 24.2 BOE. On the other hand, LPG consumption increased from 5.9 million BOE to 23.4 million BOE, and natural gas consumption increased from 81 thousand BOE to 120 thousand BOE.

Industry sector final energy consumption was 347.8 million BOE in 2009, increasing at an average annual rate of 2.4 per cent from 2000-2009. Coal use in industry notably increased from 36.0 million BOE to 820 million BOE. Natural gas supply to industry had been at an average of around 85.0 million BOE for most of the past decade however consumption increased to 117.5 million BOE in 2009, mitigation of supply constraints explains this increase. Refinery product use in industry is on a slight declining trend however remains significant at a total of 75.8 million BOE in 2009. Transport sector final energy consumption was 226.5 million BOE in 2009,

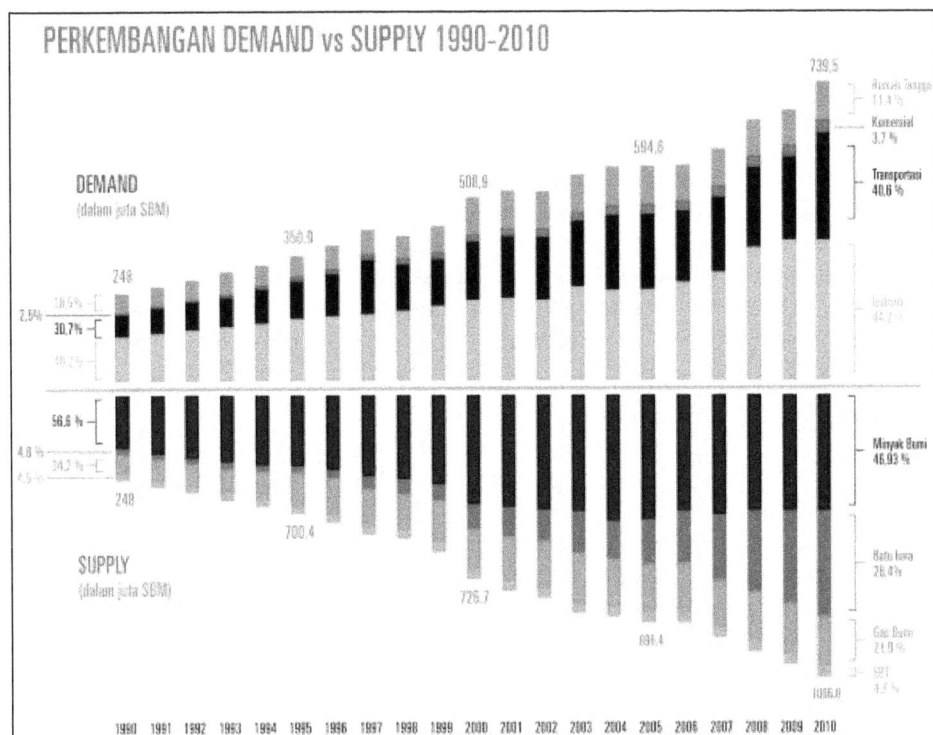

**Figure 2.2: Development of Primary Energy Demand and
Supply in Indonesia (MEMR, 2010)**

increasing at an average annual rate of 5.8 per cent from 2000 – 2009 Most notable increase of final energy consumption was for aviation turbine fuel from 7.1 million BOE in 2000 to 16.2 million BOE, and RON 88 gasoline from 70.3 million BOE in 2000 to 125.4 million BOE, in 2009. Consumption of industrial diesel oil (IDO) and fuel oil declined significantly, from combined 0.82 million BOE to 0.19 million BOE. Though relatively small in absolute terms, electricity use in transport increased by 152 per cent.

All of this growth is surely accompanied by the increase in energy demand due to the increasing number of homes, factories, and commercial and industrial buildings. If we assume that demand for electricity will grow in average 7 per cent per year for the next 30 years, then electricity consumption will significantly increase, for example in the household sector, consumption will increase from 21.52 GWh in 2000 to around 444.53 GWh in 2030. There are four main sectors of energy users, namely household, commercial, industrial and transportation sector.

Figure 2.2 also shows that, the largest energy user in Indonesia is the industrial sector with a share of 44.2 per cent. Next largest consumption is the transportation sector with 40.6 per cent, followed by the household sector with 11.4 per cent and the commercial sector with 3.7 per cent. Until now, the primary sources of energy still come from fossil fuels, with 46.9 per cent from oil, 26.4 per cent from coal, and 21.9 per cent from natural gas. Hydro (water) power and other renewable energy only make up about 4.8 per cent from the total of utilized energy resources.

3.3 Environmental Impacts

All forms of energy production and use have an impact on the environment. For example, our use of non-renewable fossil fuel energy contributes to greenhouse gas and particulate emissions, which have negative impacts on the environment and human health. The negative impacts of the energy production to the environment can be in the form of solid waste, liquid waste, and air pollution. Air pollution from burning fossil fuels can be emission of particles, VHC, SO_2, and NOx that have impact to local environment, and carbon dioxide (CO_2) with global implications.

Units used to indicate the magnitude of emission reductions is t-CO_2, so if we reduce 1 ton of greenhouse gases (other than CO_2), then the result is multiplied by its power than CO_2. CO_2 emissions represent the largest part of the GHG emissions. Forecast of CO_2 emissions in Indonesia for case R30 (low GDP growth scenario and oil prices $ 30/barrel) is shown in Figure 2.3.

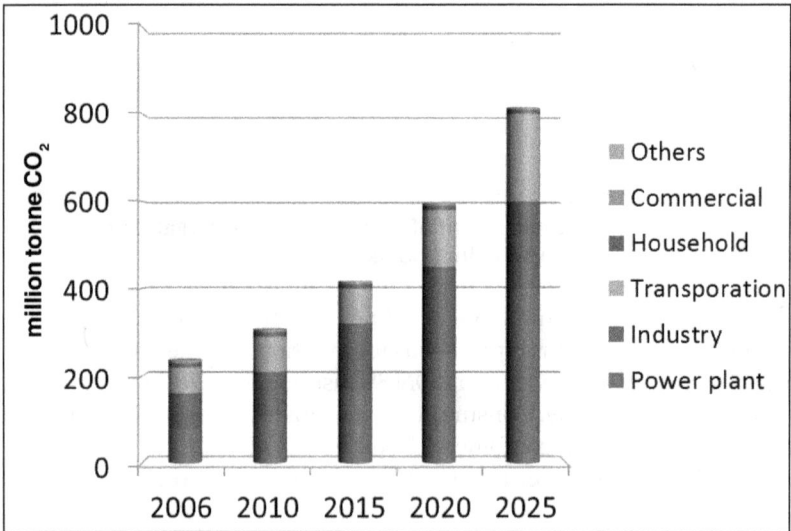

Figure 2.3: Prediction of CO_2 Emission in Indonesia (Energy Outlook, 2013)

3.4 Energy Policy and Regulation Framework

The National Energy Council (DEN) established a new National Energy Policy (KEN), called the National Energy Policy in 2006. The aim of this policy is to achieve energy elasticity to GDP of less than one by year 2025, and realize an optimum primary energy consumption mix in 2025, with shares of oil become less than 20 per cent, natural gas is more than 30 per cent, coal more than 33 per cent, bio-fuels more than 5 per cent, renewable energy and other energy including nuclear more than 10 per cent, and liquefied coal more than 2 per cent.

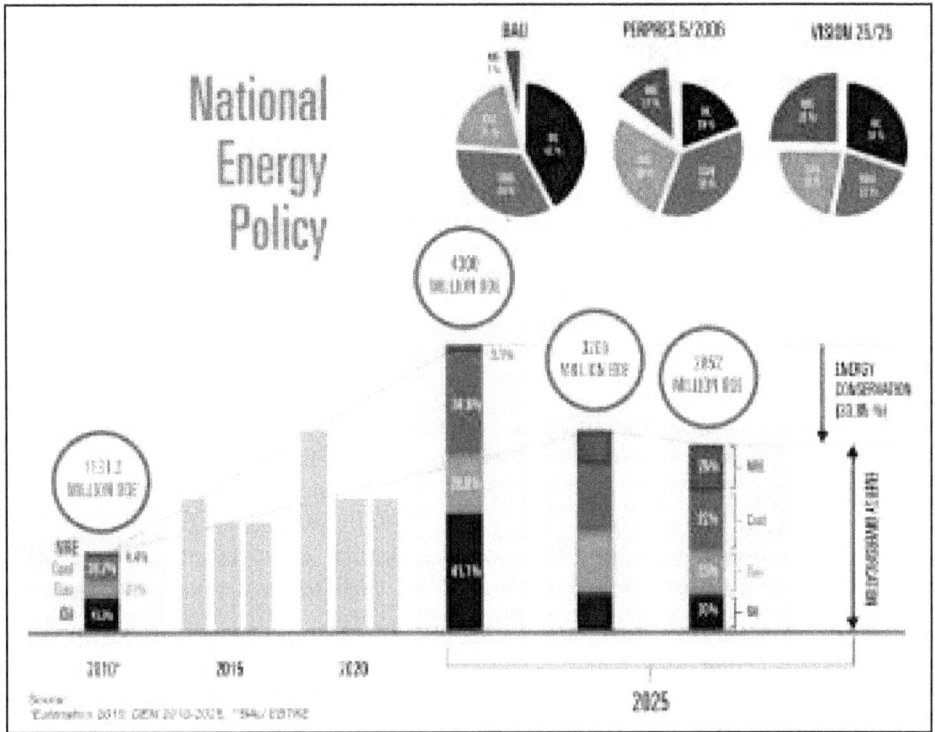

Figure 2.4: The National Energy Policy – Indonesia (MEMR, 2012)

To be more effective and efficient in the management of energy resources, the government has a strategy called Vision 25/25 (Figure 2.4), which has enacted in 2012, and outlined commitments to increase utilization of renewable energy to 25 per cent by 2025, and reducing energy demand by 33.85 per cent from business as usual scenario in 2025. This vision is pushing the target above the 17 per cent commitment set by the previous Presidential Regulation No. 5/2006 on National Energy Policy. Currently, renewable energy only makes up 4 per cent of the total energy resources being utilized. The current energy demand is 1,131 million TOE and in 2025 is expected to be 4,300 million TOE. With a variety of efforts, energy demand could be reduced to 2,852 million TOE. The government's main policy includes:

1. Conservation of Energy to improve the efficiency of the use and utilization of energy (Demand Side)

2. Diversification of Energy to increase the share of renewable energy in national energy mix (Supply Side)

The details of the energy programems and targets of the National Energy Policy are elaborated in the Blue Print - National Energy Management 2005 to 2025. Indonesia's 2006 Energy Policy expects the combined share of renewable energy in the overall energy mix in 2025 to have exceeded 17 per cent. The policy has special emphasis on enhancing the share of bio-fuels. Renewable energy and other energy (as in the list above) is expected to be made up of at least a 5 per cent geothermal share and a combined share of biomass, hydropower, solar, and wind power to make up the remainder to 10 per cent by 2025.

National Energy Conservation Master Plan (RIKEN, 2005) aims to achieve Indonesia's energy saving potential through energy efficiency and conservation (EE and C) measures, and so avoid wasteful energy use in the country. It also aims to reduce energy intensity by around 1 per cent/year, on average, until 2025. Fiscal incentives (tax reductions and soft loans) together with other instruments such as training and educational programmes as well as energy audits are used to implement the plan.

In 2007, Indonesia enacted the Energy Law (Law No. 30/2007). The Law elucidates principles for the utilization of energy resources and final energy use, security of supply, energy conservation and protection of the environment with regard to energy use, pricing of energy, and international cooperation. The Energy Law sets out the content of the National Energy Policy (KEN, Kebijakan Energi Nasional, 2006); the roles and responsibilities of the central government and regional governments in planning, policy and regulation; development priorities for energy research and development, and the role of enterprises.

The Geothermal Law (No. 27/2003), gives powers to regional governments to develop geothermal energy, in particular in respect of licensing, allowing investors to deal directly with regional governments and requiring projects to be competitively tendered. It also provides incentives for investment by establishing long-term licenses for land use (more than 30 years) and a regulated price for geothermal energy.

The Electricity Law (No. 30/2009), secures sustainable energy supplies, promotes conservation and use of renewable energy resources. The regulation was issued by the Ministry of Energy and Mineral Resources (MEMR) and referred to as purchasing price by PT PLN (Persero) of generated electricity from small and medium scale renewable energy power plant or excess power. The aim of this ministerial regulation is to enhance the electricity generated by small and medium scale of renewable energy power plant or excess power to be purchased by the state owned company, regional owned company, cooperatives. The law does not make provisions for a feed-in tariff, but does provide for differing tariffs in different regions, more accurately reflecting the cost of supply. For rural development, the government set a bold electrification target of 90 per cent by 2020. Purchase of electricity from geothermal plants (Regulation 02/2011) of the MEMR regulates the power purchase tariff of electricity from geothermal sources. It assigns the PLN to purchase electricity from geothermal plants inside at a maximum price of 0.97/kWh.

3.5 Renewable Energy Development

Renewable energy currently plays a minor role in Indonesia's energy supply, however the Government of Indonesia (GOI) is seeking an increase in renewable generation to 17 per cent of the total energy consumption by 2025. Current estimates place total renewable energy capacity at 2.9GW (on-grid) and 3.2GW (off-grid). Given a total renewable energy potential within the country of 163.3GW, this amounts to an undeveloped renewable potential of 96 per cent. The Government is aiming for 10.1GW of new capacity comprising of 3.9 GW of geothermal, and 1.2 GW of hydropower by 2014. Indonesia has introduced a Climate Change Trust Fund to ensure that climate change is integrated into the development planning. The Government is seeking to become a regional leader on climate change mitigation and in 2009 voluntarily committed to reduce greenhouse gas emissions by 26 per cent by 2020. The US$400 million Clean Technology Fund was created to promote renewable energy, improve electrification rates and enhance energy efficiency within Indonesia. The fund's predominant focus is on developing Indonesia's large-scaled geothermal projects. The Directorate General of Renewable Energy and Energy Conservation, which sits under the MEMR, is the chief regulator in the renewable energy area (DLA-Piper, 2013).

Biomass Energy/Bio-fuels

Indonesia has a large potential for biomass energy. Rice residues, sugar, rubber and palm oil all provide biomass electricity, but the most promising commercial application for biomass is likely to be cogeneration in agribusiness. The current installed capacity is 445 MW, whilst the whole potential of biomass from forestry, agriculture and estates is equivalent to 50,000 MW. Palm oil is the main biodiesel feedstock in Indonesia. In 2008, the economy produced 20 Mt, making Indonesia a leading producer and the second-largest exporter of palm oil. It was the world's 11th largest producer of bio-fuels (fuel ethanol and biodiesel) in 2010 (Reegle, 2012). Palm oil, is now mainly used in local bio-fuel development. The international price for palm oil and the higher value placed on food has meant that most palm oil is expected or used in food production.

Additional restrictions to development include a lack of incentives and subsidies for fossil fuels; incomplete data on land ownership; and a lack of bio-fuel processing ability. As a result, bio-fuel development in Indonesia remains sluggish compared to that of other countries in the region. But several companies have begun using fast-growing crops, such as cassava, *jatropha*, or sweet sorghum, for bio-fuel production. The Ministry of Agriculture is preparing additional land for growing these high-yield feedstocks to meet the country's bio-fuel production goals of 5.57 million kiloliters of biodiesel and 3.77 million kiloliters of bioethanol. The introduction of these crops will likely to accelerate the production of bio-fuels. New areas planted with *Jatropha* are expected to total 1.69 million hectares in 2010. Indonesia's biodiesel blend production capacity in 2009 was 2,865 million litres (ML) and was estimated to be 4,680 ML in 2010, far exceeding the volumes needed to fulfil the mandates in those years (Reegle, 2012).

Solar Energy

Indonesia is one of tropical countries that are considered to develop solar energy systems due to its location the country has a lot of abundant solar energy. To utilize such potential of solar energy, two technologies have been applied, namely thermal solar energy and photovoltaic solar energy. Photovoltaic (PV) systems have emerged to be the cost effective and a suitable approach for rural electrification in the rural areas of the developing countries.

Solar projects have been implemented in Indonesia since 1979, including solar home system (SHS) and street lightning system (SLS). The projects have been implemented in Sukatani village, which is the first village to use PV system to generate electricity. Many PV systems have been installed in different parts of Indonesia during the last decade. Since 2005 the application of PV systems for rural electrification program has increased significantly; as distributed system (SHS) and centralized (hybrid) system (PV, battery, diesel, or wind systems). Current installed capacity of solar PV is estimated to be about 15 MWp, which is mostly used for rural electrification in rural areas. Average daily insolation ranges from 4.5 to 5.1 kWh/m² across the country, with a monthly deviation of 9 per cent, indicating a good potential for the widespread use of solar energy.

In 2011, BPPT has developed a Smart Micro Grid in the island of Sumba, which is the first and only one in Indonesia today. This is done with the consideration that the island has a variety of potential renewable energy resources, especially solar energy, wind energy, water energy, biogas energy (from manure) and others. The potential of renewable energy resources is very likely used to provide electricity to the electrical load on the island of Sumba which has been dominated by the use of diesel generators. Figure 2.6 shows the PV power Smart Micro Grid, is located in the village of Billa Cenge, Southwest Sumba, integrating capacity of 500 kWp PV plant, two diesel power stations (in Waikabubak, 7 units and Waitabula, 4 units), and three units of mini-hydro plants in the village Lokomboro. The Smart Grid Demo Sumba is also equipped with a storage capacity of 500kWh/day, which serves as a compensator to anticipate fluctuations in power output so that penetration of PV plants can be optimized.

To accelerate development of PV application in Indonesia, the MEMR recently issued Regulation No. 17/2013, regarding to Solar Power Feed-in-Tariff and New Solar Power Purchase Procedures in Indonesia. The regulation is to stipulate among other things: (i) new procedures for purchase of power from solar photovoltaic power projects in Indonesia which require developers to bid in capacity quota tenders; and (ii) feed-in-tariff for solar photovoltaic power projects at the cap of US$0.25/kWh, or US$0.30/kWh if the photovoltaic module contains 40 per cent or more local components.

Wind Energy

Potential for wind energy in Indonesia is limited because of the lack of wind along the equator. Average wind speeds across the country range from 3-6 m/s, indicating the potential capacity for wind energy of 9,190 MW mainly from the higher wind-speed regions. However, the windiest regions tend to be the less populated, eastern islands, which lack a transmission infrastructure capable of sustaining large wind farms. Wind power opportunities are thus limited to small or medium-sized projects requiring lower wind speeds. Government plans therefore include the development of small and medium-scale wind energy systems, with a view to having a total installed wind capacity of 970 MW by 2025. To date, only a few small-scale wind farms have been attempted, and they account for only 1.6 MW of installed capacity. In the future, offshore wind is more likely to provide investment opportunities due to the lengthy coastlines and consistent ocean breezes.

The implementation of isolated wind energy systems typically in remote area/ location or islands, and they are frequently installed as part of R&D project. Several area along the coastal/shore of northern and southern part of Java Island, eastern part of Madura island, south and north Sulawesi island, east Lombok island etc, have applied wind turbine for electricity generation. Mainly used for stand-alone system and hybrid. Up to present, the largest wind power generation facility is installed at Nusa Penida-Bali, for the total capacity of 735kW, and followed by the total of 540kW wind power installation at Sangihe-North Sulawesi and Selayar – South Sulawesi. On going project to develop wind farm to setting up around 300 MW install capacity for near future at several sites at java island (Bantul, Sukabumi, Lebak), South Sulawesi (Jeneponto and Sidrap), East Nusa Tenggara (Oelbubuk).

The **Agency for the Assessment and Application Technology** through the Energy Technology Center in collaboration with the United Nations Development Programme (UNDP) is conducting the *Wind Hybrid Power Generation Market Development Initiatives Project* (*WHyPGen*). This GEF funded project is aimed to promote the development of on-grid wind power generation, especially through the combination of other energy sources such as solar, hydro or diesel, aiming to increase the use of renewable source in power generation and reduce carbon emission in the country. The program is expected to improve electricity production from wind power generation up to 19 GWh per year and reduce carbon emission by 16.071 metric tons per year by 2015.

Hydropower

Indonesia has an estimated hydropower potential of 75,670 MW. However, roughly 35,000 MW of Indonesia's hydropower potential is located in the province of Papua, which is located far from demand centre. The country has not built a large hydropower facility, and the current installed capacity is 4,264 MW. Additionally, 500 MW of mini-hydro potential have been identified, and 84 MW of which has been utilized to support rural electrification. Currently, only 6 per cent of Indonesian hydropower potential has been utilized, indicating the possibility for considerable development in the sector for the future. To provide information related on mini-micro hydro development as well as networking among all stakeholders, the MEMR has established the Mini-Micro Hydro Clearing House.

Geothermal Energy

Indonesia is home to 40 per cent of the world's known geothermal resource, which is the largest geothermal potential in the world, with the total geothermal potential of the country is roughly 28,100 MW. Current installed geothermal capacity is roughly 1,189 MW. The largest reserves of geothermal energy are located in the western part of Indonesia where energy demand is highest: Sumatra, Java and Bali. North Sulawesi makes the most advanced use of geothermal energy for electricity needs: approximately 40 percent of total electricity demand is supplied by geothermal energy.

In 2010, Indonesia was the third largest geothermal generator in the world, producing 1.2 GW. The further development of the country's substantial geothermal capacity has slowed since 2000, and currently, only 4.2 per cent of the country's total geothermal potential is being exploited. Government plans for the development of the energy sector include expanding the installed geothermal capacity of the country to 5,000 MW by 2014, rising to 9,500 MW (5 per cent of the national electricity) by 2025, according to the 2005-2025 National Energy Blueprint. The second phase of the Fast Track Program includes additional geothermal capacity of nearly 4 GW by 2014, most of which will be operated by independent power producers.

To reach this target, the government has divided into 2 scale of usage, large scale (high enthalpy) from large scale power plants, and small scale utilization (low to medium enthalpy) which includes the generator model currently being developed by the BPPT Geothermal Unit. The unavailability of a tried and proven technology has forced BPPT to innovate and develop technology which may be used to develop small scale geothermal potentials for electricity generation. This energy may then be used to power tools during the development of geothermal field. Development of binary cycle power plants in Lahendong, North Sulawesi began in the late 90's. As this project is carried out in collaboration with France, all of the components are still imported. This is different from current situations where the vast majority of equipments are locally made. Binary cycle power plants are currently in a pilot project scheme, with future plans for an extra 1MW which is divided in to several phases, explains the Director of Central Conversion and Conservation.

BPPT are currently focusing on developing small scale geothermal power plant in Kamojang, Garut, West Java. The geothermal power plant has capacity of 3 MW, with technology of condensing system. This prototype will be the only small scale geothermal power plant in Indonesia, which only requires a small investment of minimal but sufficient for the purposes of small industries and household. In addition, the power plant is fully designed by BPPT and will use a variety of local components, including turbines and generators. For turbines, BPPT took Nusantara Turbin Propulsi (NTP), an IPTN subsidiary, while for generators BPPT was in cooperation with PT Pindad.

3.6 Energy Efficiency

Indonesia's total primary energy consumption grew by nearly 30 per cent between 1999 and 2008. Total final energy consumption (TFEC) was 948,112.45 thousand BOE in 2009. Oil continues to account for the most significant share of Indonesia's energy mix, at 51.9 per cent, followed by natural gas at 18.4 per cent, coal at 12.9 per cent, electricity at 12.8 per cent and LPG at 3.9 per cent. The share of TFEC by sector in 2009 was 40.86 per cent for industry, 36.8 per cent for transport, 13.26 per cent for households, 4.73 per cent for commercial and 4.28 per cent for other sectors. There is a need for improved energy efficiency in industry, and limiting petroleum consumption in the transport sector.

There is significant potential for energy savings. A move away from general subsidies on fuel, whilst continuing to assist those who would otherwise face energy poverty, could help promote more efficient use of energy, especially oil. Indonesia has a range of options for reducing the carbon content of its energy supply. These include increased development of its hydro-electric, geothermal and wind resources, and more efficient thermal electricity generating plants.

The efficiency of the power sector tends to fluctuate and has decreased slightly since 1990. It stood at 38 per cent in 2009, compared with 40 per cent in 1990. The reduction of the share of hydroelectricity and the development of coal power plants explain the deterioration of the ratio. In 2009, thermal power plants had an efficiency rate of 35 per cent; that rate has remained relatively stable since 1990. The limited development of more efficient technologies, such as gas combined cycles and cogeneration, did not permit an increase in the ratio for thermal power generation. The rate for T and D losses in the Indonesian grid is above 10 per cent of the distributed volumes, which is above the world average of 9 per cent. Those losses tend to fluctuate significantly and have decreased since 2005.

Industrial energy consumption has soared by 6 per cent/year since 1990, which is faster than the economic growth rate of 4.5 per cent/year. Industrial energy intensity increased by 0.7 per cent/year between 1990 and 2009. Since 2000, it has increased at the even faster pace of 1.2 per cent/year. In 2005, the RIKEN identified the following sector of energy saving potentials: 15-30 per cent in industry; 25 per cent in commercial buildings for electricity; and 10-30 per cent in the households sector. Energy Intensity reduction: 1 per cent/yr until 2025, through fiscal incentives, educational programs and audits. The Partnership Program on Energy Conservation – Government support in return for efficiency commitments from companies, 20 per cent efficiency improvement targets.

3.6.1 Energy Audit and Conservation

To support the government policy on energy efficiency, the Energy Technology Center (B2TE-BPPT) has developed some programs on energy audits, energy monitoring system for buildings, and cogeneration technology assessment programs. The energy audits were performed either under government program or commercial based contracts. The energy audits activity was started since 1989 until now. Under Government to Government contract we have done for industrial sector energy audit in food and beverages and for hotel sector. While under commercial based we have performed several energy audits for commercial buildings, sugar industries, cement plants, steel industries, and in power plant sector. Energy Audits are activities to measure, analyze and evaluate the system of energy usage in energy user (idustries, buildings and commercial sectors). This audit will be identified energy savings opportunities. In the management energy system, energy audit is an important part to measure system performance. In this year, several partners have been cooperated with B2TE-BPPT in Energy Management and Audit, such as : PT Cussons, PT Great Giant Pine apple, Hotel Parklane, PTPN IV, Ministry of ESDM, Ministry of Indutry, Cluster Building of BPPT Serpong.

The long experiences in energy audits have initiated our program on energy usage monitoring that could help energy user to control their program on energy conservation. Our first product on energy monitoring was SIME (Sistem Informasi Manajemen Energi/Management Energy Information System) in 2012. Figure 2.5

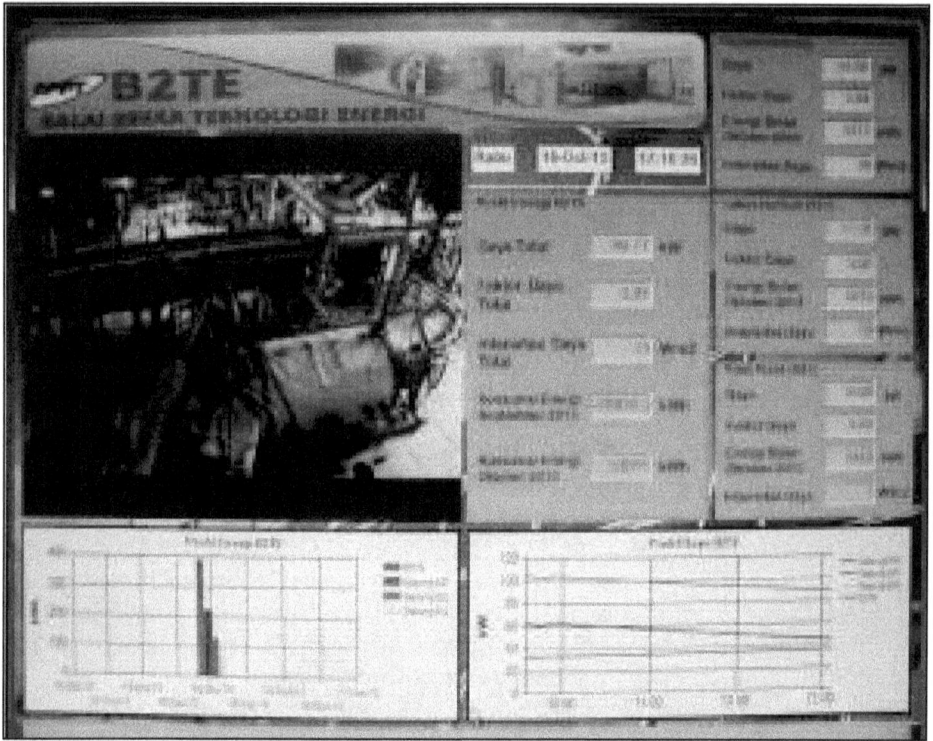

Figure 2.5: Energy Management Information System

show the SIME used to monitor energy usage for buildings and has been implemented in our building of B2TE and others buildings are in progress. As a pilot project, the system has been applied in B2TE 620-BPPT building at Serpong. Application of the system in industrial sectors, will provide tangible benefits. Industry can get ISO 50001 accreditation standards appropriately, as well as a real step in the application of government regulations in PP 70, 2009 and No. 14 PERMEN ESDM, 2012. Future development for this monitoring system will be developed to other implementation for industrial sectors.

3.6.2 Cogeneration Technology

Cogeneration Technology, which also popularly called combined heat and power generation is a system of generation to produce electricity and thermal energy simultaneously. In the assessment of cogeneration technology to improve energy efficiency in industrial sectors and commercial buildings, our institution has assessed microturbine application for small scale cogeneration system under the program of MCTAP (Microturbine Cogeneration Technology Application Program) under funding from GEF and monitored by UNDP. The MCTAP promotes and diseminates the application of microturbine cogeneration technology (MCT). The MCT is a small scale turbine generation (under 1MW/unit) that produce energy more efficient and less emission. MCT application only produce CO_2 emission of about 0.16kg/kWh.

One of the MCTAP demo units, MCT C65 (65 kW electrical capacities) application at hotel area, provide additional electricity supply of about 51 kW to the grid, reducing the burden of steam calorifier 15-70 liters/minute and reduce the burden on existing boilers by 89 kW(th). Hot water generated from MCT can supply 40-60 per cent of the hotel guest rooms.

B2TE-BPPT also studies the potential for cogeneration applications in the industrial and national power plants. For the industrial sector, cogeneration technology has the potential energy savings up to 13.8 thousand GWh per year, equivalent to reduce of CO_2 emissions by around 10.7 million tonnes per year. Meanwhile, if applied to national (PLN) power plants, cogeneration technology has potential energy savings up to 13 thousand GWh per year, equivalent to reduce of CO_2 emissions of about 10.1 million tons per year.

4. CONCLUSIONS

Energy consumption in Indonesia is still dominated by fossil energy (oil, gas, and coal), which is limited and harm to the environment. Meanwhile, development of new and renewable energy (NRE) and energy efficiency in Indonesia is still relatively slow. Indonesia is one of tropical countries, having a lot of abundant renewable energy resources, such as solar energy, hydro, geothermal, biomass, etc. Besides, Indonesia has a significant potential for energy savings in the following sectors: 15-30 per cent in industry; 25 per cent in commercial buildings for electricity; and 10-30 per cent in the households. A move away from general subsidies on fuel, whilst continuing to assist those who would otherwise face energy poverty, could help promote more efficient use of energy, especially oil. Some efforts should be taken to increase the use of renewable energy and energy efficiency.

5. ACKNOWLEDGEMENTS

Herewith, we would like to convey our gratitude to (1) the Society of Energy Engineers and Managers (SEEM), Trivandrum, India (in association with Christ University Faculty of Engineering, Bengaluru) jointly with the Centre for Science and Technology of Non-Aligned and Other Developing Countries (NAM S&T Centre), (2) the Ministry of Research and Technology of Republic of Indonesia and to (3) Agency for the Assessment and Application of Technology for all their support of our participation in a 3 - days 'International Workshop on 'Sustainable Energy for All: Transforming Commitments to Action' during 22-24 February, 2014 at Christ University, Bengaluru, India.

REFERENCES

1. Sugiono, A., Adhi Dharma, M.Sidik B, Adiarso, 2013. Indonesia Energy Outlook 2013. BPPT. Jakarta, Indonesia.

2. MEMR. 2006. Blueprint Pengelolaan Energi Nasional 2005–2025. Director General for Electricity and Energy Utilization Ministry of Energy and Mineral Resources, Republic of Indonesia. Jakarta, Indonesia.

3. Hanan Nugroho. 2005. Financing Renewable Energy Utilization in Indonesia: Notes, Proceeding of the World Renewable Energy Regional Congress and Exhibition (WRERCE) 2005. Jakarta, Indonesia.

4. Reegle info. 2012. Energy Profile Indonesia. Clean Energy Information portal

5. http: www.reegle.info/countries/indonesia-energy-profile/ID

6. Stephen Webb, 2013. Indonesia: Renewable energy in the Asia Pacific: a legal overview. 3rd edition. DLA-piper, Australia.

7. MEMR. 2010., Handbook of Energy and Economic Statistics of Indonesia, Center for Data and Information on Energy and Mineral Resources, Ministry of Energy and Mineral Resources, Jakarta, Indonesia.

8. MEMR. 2008, Indonesia's Renewable Energy Potential. Ministry of Energy and Mineral Resources. Jakarta, Indonesia.

Chapter 3

Transforming Sustainable Energy Commitments into Actions in Mauritius

Vishwamitra Oree

*Department of Electrical and Electronic Engineering,
Faculty of Engineering, University of Mauritius
E-mail: v.oree@uom.ac.mu*

ABSTRACT

Rising energy demands along with increasing cost of fossil fuels and their associated environmental threats have spawned a new global energy paradigm. Many countries are advocating an energy transition aiming at reducing the dependence on conventional fuels while expanding the integration of sustainable energy sources. Recently, renewable energy sources and energy efficiency have been the leading drivers of this transition, catalysed by a combination of factors which include progress in technology, innovation, improved cost-effectiveness and government policies. With its energy requirements highly reliant on fossil fuels, Mauritius has also made a strong commitment to build a sustainable energy future by endorsing the "Maurice Ile Durable". The latter constitutes a new sustainable vision that pervades many different sectors of the economy, with a view to guide national development. In the energy sector, the government has set a target to achieve 35 per cent self-sufficiency in terms of electricity production by the year 2025 and 10 per cent reduction in the energy consumption of non-residential buildings by 2020. Several initiatives have already been implemented while others are in the pipeline to meet the objectives set. Projects involving photovoltaic and wind turbine technologies are currently being developed to harness the significant solar and wind energy resources available in the country. It is expected that a total renewable energy capacity exceeding 60 MW will be connected to the grid when these projects will be fully operational in 2015. An Energy Efficiency Management Office was set up in December 2011 as a nodal agency for the development and implementation of energy efficiency measures in all sectors of the economy.

Keywords: Energy efficiency, Greenhouse gas emissions, Small island, Photovoltaic, Wind power.

1. INTRODUCTION

1.1 Global Energy Situation

Energy underpins development and economic growth. Since the dawn of the Industrial Revolution, global energy consumption has soared inexorably, catalysed by population growth and economic development. An overwhelming share of the energy demand has been met by burning fossil fuels as illustrated in Figure 3.1. Between 1970 and 2005, global final energy consumption nearly doubled, with an annual growth of 2 per cent until 2002 and 4.5 per cent for the remaining three years (Randolph and Masters, 2009). The unremitting exploitation of the non-renewable fossil fuel reserves of the planet for energy-related activities has contributed to about 61 per cent of the total anthropogenic greenhouse gas (GHG) emissions (Baumert *et al.*, 2005). Moreover, the volatile nature of oil prices put at risk economies that are heavily dependent on fossil fuels. An affordable and environmental-friendly system of energy supply, which meets the energy demand cost-effectively and sustainably, is therefore essential in ensuring the energy security of a country while mitigating global climate change.

Renewable energy (RE) sources have emerged as a credible alternative to conventional fuels since the beginning of this century. Their growth has been spurred by a string of governmental actions that includes compelling measures such as carbon taxes on greenhouse gas emissions, feed-in tariffs that guarantee profitable wholesale prices and tax credits for generation of electricity through green sources. Many

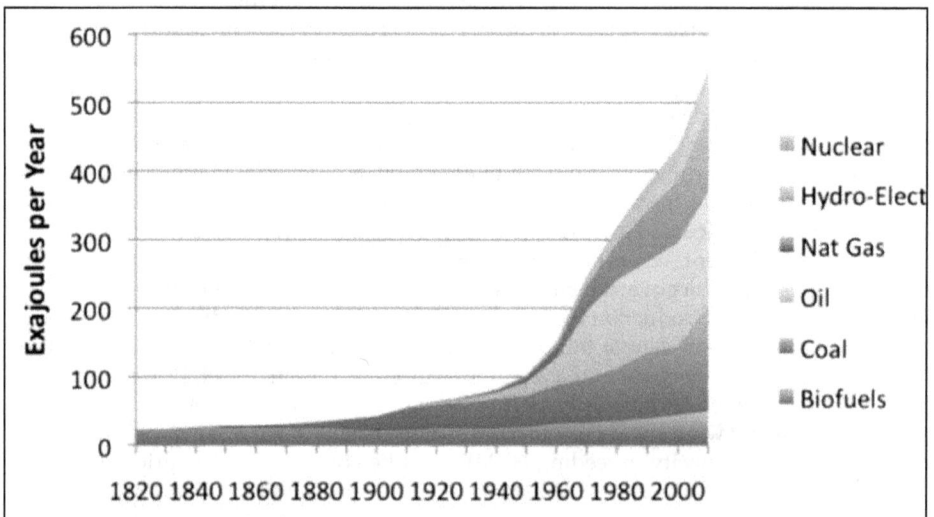

Figure 3.1: History of the World Energy Consumption (Smil, 2010)

countries have set national goals for RE deployment in terms of the share of the total energy mix in the future. It has been reported that 19.46 per cent of the world's total electricity was generated by RE sources in 2009. Out of this share, 83 per cent was attributable to hydro while biomass, solar and wind contributed about 15 per cent (IEA, 2011). Another crucial element in ensuring a reliable and low-carbon energy portfolio is energy efficiency. A huge amount of energy is wasted through inefficiencies in the energy generation, transmission, distribution and consumption processes. Energy efficiency not only curtails the energy demand growth and carbon emissions but also represents a very cost-effective solution to the global energy problems both in the short-term and the long-term. Improved technologies and innovation have led to much more energy efficient buildings, vehicles and appliances. It is estimated that energy efficiency represents a GHG reduction potential of about 40 per cent at an affordable cost (McKinsey and Company, 2009).

Most countries today are experiencing the modern energy transition even though it is occurring at vastly different rates across various world regions. Traditionally, developed countries have been overwhelmingly responsible for the sustained growth in energy demand. However, it is anticipated that in the near future, the bulk of the energy demand growth will emanate from developing countries, bolstered by population increase and improved standards of living. It is forecasted that by the year 2035, almost 95 per cent of the total growth in energy consumption will come from the non-OECD countries (BP, 2014). The global energy scenario is even more important for Small Island Developing States (SIDS), which are most vulnerable to the impacts of climate change on their key economic sectors like tourism, agriculture and fisheries. Moreover, most SIDS do not have any indigenous reserve of fossil fuels and are largely dependent on fuel imports for their energy requirements. Their economies are therefore exposed to fluctuations in the prices of fossil fuels. In order to shift from a position of vulnerability and dependence to one of resilience and sustainability, SIDS must endorse the global energy transition.

1.2 Energy Situation in Mauritius

Mauritius, a small island of 1870 km^2 located in the South West of the Indian Ocean, has recently initiated such a transition. The Republic of Mauritius consists of the main island of Mauritius and other smaller islands including Rodrigues having an area of 108 km^2. The total population stands at about 1.3 million. The island of Mauritius depends heavily on foreign oil imports for a major share of its energy requirements. In 2011, 83.8 per cent of the total primary energy requirements were met from petroleum products and coal with the rest being supplied by a mix of renewable resources consisting mostly of hydro, landfill gas and bagasse (CSO, 2012). Thus, the stake of energy imports in the total import expenditure of the island is substantial and on the rise as illustrated in Figure 3.2.

Figure 3.2 also underlines the vulnerability of the Mauritian economy to the volatility of oil prices. As the latter reached at all-time high in July 2008, the share of fossil fuels in the imports bill increased beyond 20 per cent whereas it decreased by 4.7 per cent when the oil prices moved to unusually low values the following year. In

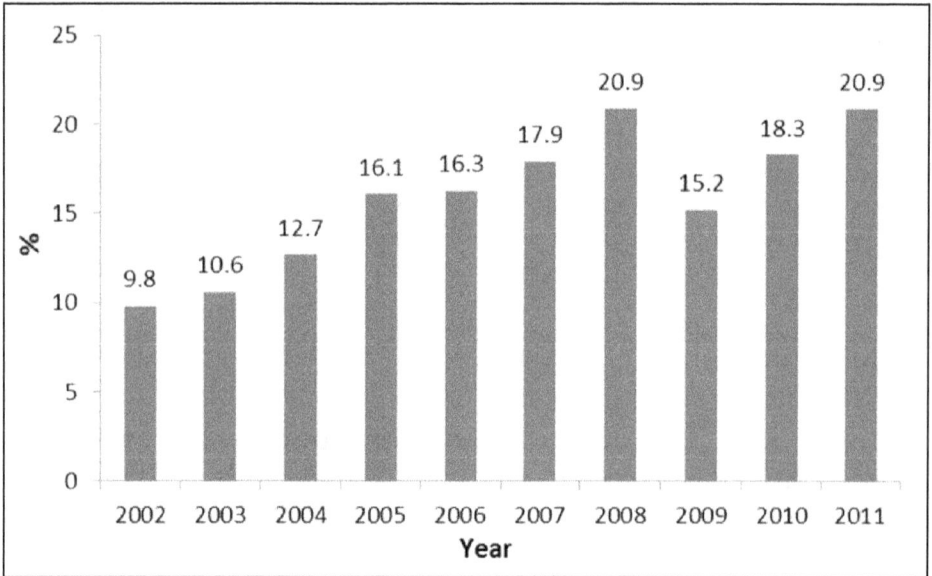

**Figure 3.2: Energy Imports as a Percentage of Total Imports Value
for Period 2002-11 (CSO, 2012)**

real monetary terms, the net cost for the imports of fossil fuels has increased about threefold in five years, from MUR 9.7 billion in 2004 to MUR 27.6 billion in 2008 (MENDU, 2010). Clearly, a business-as-usual strategy in the energy sector is economically unsustainable as the energy demand continues to grow every year. Against this backdrop, the government decided to act responsibly by proposing a comprehensive and integrated sustainable development policy through the "Maurice Ile Durable (MID)" in 2008. Literally meaning "Mauritius Sustainable Island", the policy initially focussed on integrating more RE and energy efficiency into the energy mix. This would result in a diverse energy portfolio that would meet the energy demand sustainably while mitigating energy insecurity. Subsequently, the MID concept widened to encompass all aspects of the economy, society and the environment, that are considered to be pivotal in the pursuit of a sustainable Mauritius (MESD, 2011). An MID Fund was established to financially support endeavours that abide to the sustainability concept. The Fund is partially derived from levies imposed on all imported fuels. This paper reviews the tangible actions taken by the government to translate its commitment to make Mauritius a sustainable model into reality.

2. ENERGY FORECASTS FOR MAURITIUS

The energy demand in Mauritius has been increasing since its independence in 1971. The advent of energy-intensive industries in the early 1990s, as the island shifted from a mono-crop economy, spawned a faster growth in energy demand. The growth accelerated at the start of the century, driven by the expansion of existing industries like tourism, manufacturing and commercial as well as the emergence of new economic sectors such as information and communication technologies and financial services. Concurrently, the standard of living of the population improved considerably, thus exacerbating the rise in energy demand. Planning the energy resources required to satisfy the demand over the long-term is an extremely complex exercise as much depends on many economic, demographic, technological and policy variables that are often beyond the control of a country. The national electricity utility, the Central Electricity Board (CEB), has provided a glimpse into how the electricity demand will grow during the next ten years by using a methodology based on the causal effect of the basic aforementioned influential factors (CEB, 2013). Figure 3.3 shows the forecasted electricity demand for the country if the growth rates of these factors are assumed to be comparable to those of the last decade. It is observed that the average yearly increase in electricity demand is predicted to be around 78 GWh.

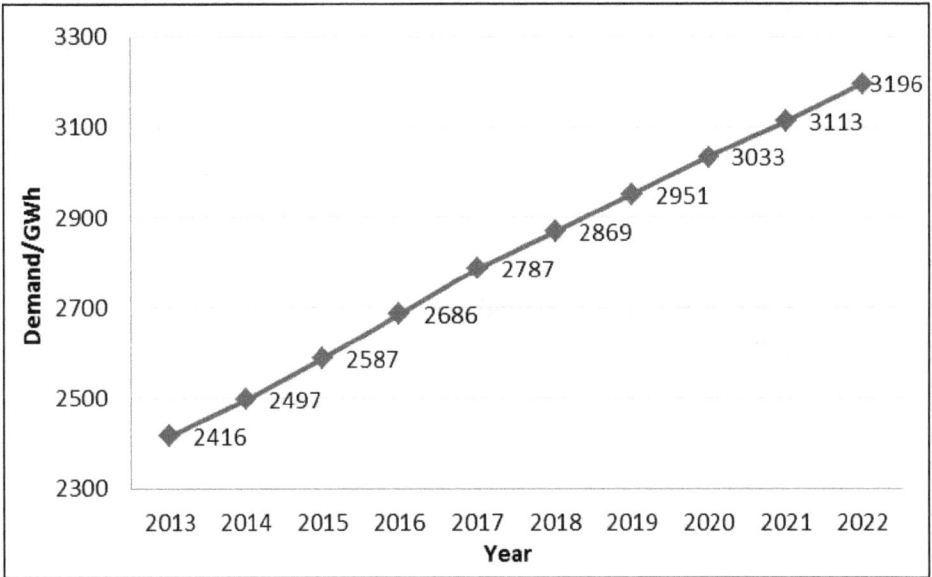

Figure 3.3: Electricity Demand Forecast for the Period 2013-2022 in Mauritius (CEB, 2013)

3. RENEWABLE ENERGY IN MAURITIUS

3.1 Renewable Energy Targets

The first step taken by the government to provide a reliable, adequate and affordable energy supply to all Mauritians was to work on an elaborate roadmap for the next fifteen years. Its objective was to steer the country successfully through the change to an infrastructure capable of meeting the energy challenges. One of the highlights of this long-term plan is to generate 35 per cent of the total electricity consumption of the island through RE sources by 2025. The figures given in Table 3.1 refer to progressive targets that have been set in terms of percentage of total electricity generation from various RE sources over the period 2015-2025. It is noted that the country intends to tap its wind, solar and biomass resources immediately. Geothermal technology is expected to be deployed as from 2020 when the technology matures and the high upfront investment costs drop. Although tidal and wave power technologies hold high potentials for the island, they have been excluded as the technologies are still in their infancy.

Table 3.1: Targeted Renewable Energy Mix for the Period 2015-2025 (MREPU, 2009)

Fuel Source		Percentage of Total Electricity Generation			
		2010	*2015*	*2020*	*2025*
Renewable	Bagasse	16 per cent	13 per cent	14 per cent	17 per cent
	Hydro	4 per cent	3 per cent	3 per cent	2 per cent
	Waste to energy	0	5 per cent	4 per cent	4 per cent
	Wind	0	2 per cent	6 per cent	8 per cent
	Solar Photovoltaic	0	1 per cent	1 per cent	2 per cent
	Geothermal	0	0	0	2 per cent
	Sub-total	**20 per cent**	**24 per cent**	**28 per cent**	**35 per cent**

3.2 Setting Up the Framework

Generating a significant fraction of total electricity requirements of a country from renewable sources is not feasible without garnering the support of the population at large in the process. In line with the Government's policy to allow new players in the electricity market and to enable small-scale electricity producers to produce their own electricity and sell any surplus generation to the national grid, a Small-Scale Distributed Generation (SSDG) project was introduced in December 2010. A Grid Code was formulated to define the technical guidelines and operational requirements for the interconnection of small-scale power producers to the low-voltage network system. The preferential feed-in tariff offered to the small-scale producers for selling electricity from their facilities to the grid is given in Table 3.2. When compared to the rate of MUR 3 to 8 charged by the CEB to consumers, it can be deduced that the proposed feed-in tariffs provide an interesting financial incentive to potential investors.

Table 3.2: Feed-in Tariff for Grid Code (CEB, 2010)

Feed-in Tariff for 15 years	WindMUR/kWh	HydroMUR/kWh	PVMUR/kWh
Micro (up to 2.5 kW)	20	15	25
Mini (2.5 to 10 kW)	15	15	20
Small (10 to 50 kW)	10	10	15

3.2 Solar Power

Table 3.1 reveals that solar energy is regarded as one of the most viable RE options for Mauritius. The solar energy potential of the island was central in justifying the decision to invest in solar energy. According to data from the Mauritius Meteorological services, the tropical island of Mauritius enjoys some 2,400 to 2,850 hours of sunshine annually depending on the location. Figure 3.4 depicts the average

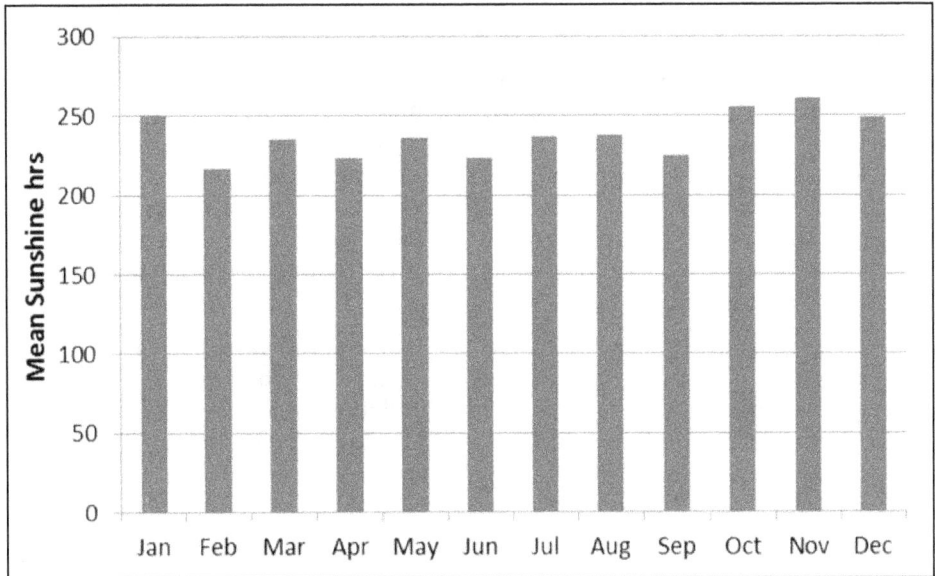

Figure 3.4: Average Monthly Total Sunshine Hours in the North of Mauritius (MMS)

daily sunshine hours measured on a monthly basis in the northern part of Mauritius. The average solar radiation over the island is 5.4 kWh/m²/day.

These promising figures spurred the Government to initiate a solar water heater programme under the MID project. During the first stage of the programme in 2008, 24,000 families were provided with a subsidy of MUR 10,000 each to help them purchase and install solar water heaters on the roof of their houses. Following the success of the first stage, the second phase of the Scheme was launched in December 2011 to support 14,000 families in procuring a solar water heater. The total budget of these two phases amounted to MUR 384.47 million. The third phase of the project started in October 2012 and is expected to equip an additional 20,000 households with solar water heaters. By the end of its third phase, the programme will enable about 17 per cent of the 330,000 households in the island to tap solar energy to get hot water in their homes.

Encouraged by the positive response received with solar water heaters, the government decided to step up its determination to harness the solar potential of the country for electricity generation from photovoltaic (PV) technology. Prior to 2010 and the advent of the Grid Code, the country had very limited experience with grid-connected PV electricity generation. The subsidies involved in the SSDG project and its associated Grid Code have motivated small independent power producers and households to invest in small-scale electricity generation through PV. Initially launched in December 2010 for residential, commercial and industrial customers with individual installed capacities of less than 50 kW, the SSDG scheme was targeting a total generation capacity of 2 MW. Given the high number of applications received from interested parties, mostly employing PV, the total capacity of the SSDG scheme was increased to 3 MW in December 2011. 200 kW out of the total capacity were reserved for the island of Rodrigues. 403 applications were received for the remaining 2.8 MW capacity allocated to Mauritius, from which 341 were retained after processing. Until recently, the installations of 77 small independent power producers representing a total capacity of about 714 kW were already commissioned. On the other hand, 55 applications were received to supply the 200 kW for Rodrigues and 35 were retained (CEB, 2013). The SSDG project has yet another scheme which caters for a total capacity of 2 MW to be generated by public, educational and charitable institutions. The applicable tariff under this scheme is the marginal cost of production of the CEB, given the nature of the targeted institutions.

In the wake of the success of the SSDG project, several larger electricity consumers and producers expressed their interests in setting up RE plants with capacities exceeding 50 kW. A medium-scale distributed generation (MSDG) scheme was set up to cater for these demands. Two Grid Codes are presently being developed to describe the technical criteria and requirements for the connection of distributed generation plants of capacities 50 kW to 200 kW and 200 kW to 2 MW respectively (CEB, 2013).To further increase the share of renewables in the electricity generation portfolio and to benchmark purchase prices of solar technology, the CEB launched a bid in March 2012 for a 10 MW Solar PV Farm, in units of 1 or 2 MW. The bids received for the project revealed that the preferred bidders agree to sell electricity to CEB at a price of Rs 5.90/kWh. An annual amount of MUR 27 million has been earmarked from the MID Fund as a financial support for the project (RM, 2013). Moreover, the CEB has signed an energy supply and purchase agreement with Sarako PV Co. Ltd. in May 2013 for the construction of a PV farm of 15 MW at Bambous, a village located in the West of Mauritius. The farm will accommodate 62,000 PV modules over a total land area of 337,668 m². Under the agreement, CEB will purchase electricity from the promoter at a price of MUR 6.06/kWh. The project will receive an annual funding of MUR 37 million under the MID Fund (RM, 2013).

3.4 Wind Power

Wind power generated in the island of Mauritius is presently negligible. In the late 1980s, wind turbines were installed in Mauritius on a pilot basis. However, they were not designed to withstand strong winds and were eventually destroyed by the fierce gusts of cyclones. No further major wind projects were initiated until the country set out to curb its dependence on fossil fuels. The island of Rodrigues took the lead by installing three wind turbines, each of capacity 60 kW, at the Trefles site in 2008. This was followed by four additional units of 275 kW each at the Grenade site to bring the total installed wind power capacity in Rodrigues to 1,280 kW. The wind turbines are all of tilted type so that they can be protected during cyclones (MREPU, 2009). In 2011, 9 per cent of the total electricity generated in Rodrigues was produced by the wind farms, representing 2.97 GWh (CEB, 2013). The commercial viability of wind power in Rodrigues motivated the government to consider its generation on a large-scale in Mauritius. Table 3.1 indicates that wind is regarded as the most promising RE resource given that its contribution to the total electricity mix is predicted to reach 8 per cent in 2025. The optimism shown for a bright wind power future hinges on the fact that just like Rodrigues, the island of Mauritius is exposed to South East trade winds for most of the year. Figure 3.5 depicts the monthly mean wind speed in the southern part of the country, indicating a good wind regime.

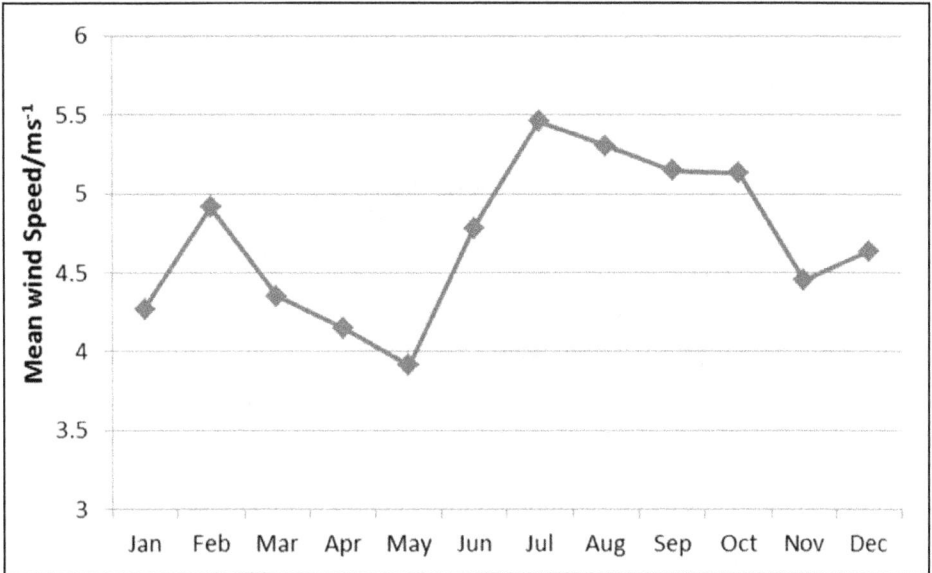

Figure 3.5: Monthly Mean Wind Speed in the South of Mauritius (MMS)

To kick-start wind power development in Mauritius, CEB signed two energy supply and purchase agreements with foreign RE experts, Suzlon Padgreen and Aerowatt, for wind farms of capacities 29.4 MW and 9 MW in the centre and North-East of the island respectively. Some 77 GWh of wind-generated electricity is expected to be coupled to the grid on an annual basis when the two projects are completed (CEB, 2013).

3.5 Other Renewable Sources

The government is also planning to exploit the 420,000 tonnes of solid waste produced in Mauritius annually to generate electricity. Presently, solid waste is disposed of at the unique landfill of the island, situated in the village of Mare Chicose, after being collected and compacted at four transfer stations. Studies have been carried out to assess the potential of using the gases emanating from the landfill as a source of energy. Landfill gases consist of potent greenhouse gases, with about 50 to 55 per cent of methane and the rest being carbon dioxide (CO_2). It was estimated that the energy production potential in 2010 was 50.50 GWh (Surroop and Mohee, 2011). The encouraging results urged the government to implement a Landfill Gas to Energy plant at Mare Chicose through Sotravic Ltd, the private company responsible for the management of the wastes since the inception of the landfill. Thus, a plant of 3.3 MW capacity has been in operation since November 2011 and is expected to generate nearly 22 GWh of electricity annually. The project also enabled Mauritius to obtain United Nations Authorised Carbon Credits issued under the Kyoto Protocol for the first time (Sotravic, 2011).

3.6 Challenges Facing Renewable Energy Integration

The output of solar and wind power plants cannot be controlled and predicted as the weather conditions at a specific location at any instant affects the amount of sunlight and wind that is available. The integration of intermittent RE technologies like wind and solar to the grid therefore poses significant technical challenges. The non-controllable, variable, unpredictable and location-dependent nature of solar and wind imply that adding these resources to the electricity mix involves managing other grid operations like frequency and voltage control, expansion of energy storage, additional ramp generating capacities and load management (Perez-Arriaga, 2011). For these reasons, the CEB will assess the impact of the output from the approved solar and wind projects on the stability and security of its network once all the installations are connected to the grid. The findings of the assessment will be crucial in deciding whether further intermittent capacity can be integrated in the network (CEB, 2013).

4. ENERGY EFFICIENCY MEASURES

Besides tapping the RE resources of the island, improving energy efficiency in all sectors is the other cornerstone of the government policy to limit reliance on fossil fuels. Efficiency improvements that can decrease the energy required for various applications can go a long way in moderating the energy demand growth. One of the first initiatives undertaken following the onset of the MID policy was to target inefficient lighting technologies in the housing sector. In 2008, the CEB sold one million compact fluorescent lamps to its residential customers at a nominal price in order to foster the phasing out of the popular and energy inefficient incandescent lamps. As a result, a drop of 15 MW in the evening peak was observed. This corresponds to an annual decrease in electricity consumption of 16 GWh (CEB, 2013). Achieving energy efficiency nationwide calls for a concerted effort. In this context, numerous sensitisation campaigns have been conducted among the population at large, to raise awareness on simple energy saving ideas and their benefits.

Following the vote of an Energy Efficiency Bill in the parliament in April 2011, an Energy Efficiency Management Office (EEMO) was established in December 2011 as a nodal agency for the development and implementation of energy efficiency measures in all sectors of the economy, including the transport, industrial and household sectors. The EEMO is presently working on the formulation of strategies and enforcement of regulations for labelling of electrical appliances and new building codes. In the transport sector, a carbon tax has been introduced in 2010 on cars that have a CO_2 emission rate higher than 158 g/km. On the other hand, a rebate is granted when the vehicle has a CO_2 emission rate that is lower than the threshold.

Many schemes to further improve the energy efficiency are under consideration. One of them relates to the replacement of T8 fluorescent lamps by the less energy-hungry T5 lamps on a large-scale. Along the same lines, the possibility of promoting the use of LED light bulbs is being envisaged although their high cost is a barrier presently. The prospect of avoiding power consumption of electrical appliances in stand-by mode is interesting for residential consumers. For this purpose, the use of Smart Power Strips to intelligently control standby power consumption is being studied. Finally, the installation of smart meters will enable consumers to monitor their loads and consumption in real-time (CEB, 2013). All these initiatives can impact positively on the electricity bill of the consumer if they are supported by an effective communication campaign.

5. CONCLUSIONS

Mauritius has embarked on an ambitious long-term project to decrease its dependence on fossil fuels and cut its GHG emissions substantially. It aims at progressively increasing the autonomy of its energy mix with a view of reaching a threshold of 35 per cent by 2025. To meet this stringent target, investments in renewables constitute tangible actions on the supply side of the energy equation, while enhancing the energy efficiency represents actions on the demand side. The provision of the necessary regulatory framework, attractive feed-in tariffs and government subsidies have boosted the development of solar and wind energy. In addition to the SSDG scheme, several agreements have been signed with international companies to supply solar and wind power. The completion of all these projects will add a RE capacity of about 68.4 MW to the grid. The conversion of landfill gases to energy is already contributing a capacity of 3.3 MW to the grid. Government aims at setting up 20 MW and 10 MW capacities of wind and solar power respectively every three years to sustain the impetus. These endeavours have put Mauritius on course to achieve its objectives three years into the long-term energy strategy. However, it is acknowledged that conventional fuel backup plants and upgrade of the existing electricity network infrastructure will be necessary to cater for the intermittency of solar and wind energy as their shares increase in the overall generation mix.

ABBREVIATIONS

CEB:	Central Electricity Board
CO_2:	Carbon Dioxide
EEMO:	Energy Efficiency Management Office
GHG:	Greenhouse Gases
LED:	Light Emitting Diode
MID:	Maurice Ile Durable
MSDG:	Medium-Scale Distributed Generation
MUR:	Mauritian Rupees
OECD:	Organisation for Economic Co-operation and Development
PV:	Photovoltaic
RE:	Renewable Energy
SIDS:	Small Island Developing States
SSDG:	Small-Scale Distributed Generation

REFERENCES

1. Baumert, K.A., Herzog, T., Pershing, J. 2005: Navigating the Numbers, Greenhouse Gas Data and International Climate Policy. World Resources Institute,

Washington D.C., USA, pp. 41. (http: www.wri.org/publication/navigating-numbers. Accessed 25 January 2014)

2. BP, 2014. BP Energy Outlook 2035. (http: www.bp.com/content/dam/bp/pdf/Energy-economics/Energy-Outlook/Energy_Outlook_2035_booklet.pdf. Accessed 27 January 2014).

3. CEB (Central Electricity Board), 2010. Grid Code for Small Scale Distributed Generation.

4. CEB (Central Electricity Board), 2013. Integrated Electricity Plan 2013-2022.

5. CSO (Central Statistics Office), 2012. Digest of Energy and water statistics 2011, Republic of Mauritius.

6. IEA (International Energy Agency), 2011.World Energy Outlook 2011. OECD/IEA.

7. McKinsey and Company, 2011. Pathways to a Low-Carbon Economy: Version 2 of the Global Greenhouse Gas Abatement Cost Curve.

8. MESD (Ministry of Environment and Sustainable Development), 2011. Maurice Ile Durable Green Paper - Towards a National Policy for a Sustainable Mauritius. Republic of Mauritius.

9. MENDU (Ministry of Environment and National Development Unit), 2010. Mauritius Strategy for Implementation National Assessment Report 2010. Republic of Mauritius, pp.8.

10. MMS (Mauritius Meteorological Station) website. (http: metservice.intnet.mu/climate-services/climate-of-mauritius.php Accessed 30 January 2014).

11. MREPU (Ministry of Renewable Energy and Public Utilities), 2009. Long-Term Energy Strategy 2009 – 2025. Republic of Mauritius.

12. Perez-Arriaga, I., 2011. Managing Large Scale Penetration of Intermittent Renewables, MITEI Symposium on Managing Large-Scale Penetration of Intermittent Renewables, Cambridge, U.S.A.

13. Randolph, J. and Masters, G. M. 2009. Energy for Sustainability – Technology, Planning and Policy. Island Press, Washington D.C., USA, pp. 8.

14. RM (Republic of Mauritius), 2013. Fifth National Assembly Parliamentary Debates (Hansard), Second Session, 04 June.

15. Smil, V. 2010. Energy Transitions: History, Requirements, Prospects. Praeger Publishers, Santa Barabara, CA, USA, pp. 154.

16. Sotravic Ltd website, 2011. (http: www.sotravic.net/news/11-sefa-supports-sotravic-ltd-innovative-project.html. Accessed 01 February 2014).

17. Surroop, D. and Mohee, R., 2011. Proceedings of 2[nd]International Conference on Environmental Engineering and Applications. IACSIT Press, Singapore, pp. 237-241.

Chapter 4
Effect of Irrigation, Energy and Environmental Factors on Agricultural Productivity in Nigeria

Ezekiel Ayinde Alani

Department of Agricultural Economics,
Ladoke Akintola University of Technology, Ogbomoso, Nigeria
E-mail: aaezekiel@lautech.edu.ng

ABSTRACT

The success of agricultural production is measured by the amount of solar energy, water consumption that is captured and converted into food per unit land area as a result of manipulating plant, land, water and other resources. Irrigation schemes in developing countries especially in Sub-Sahara Africa (SSA) including Nigeria suffer from very low water use efficiency, resulting in water logging and salinity problems. Therefore, this study analyzed the problem of how urban and irrigation interests can work together and to work on how to meet future water needs of Nigeria. The main objective of this study therefore is to analyze the economic impact of water, irrigation, energy and environmental factors on agricultural productivity in Nigeria. Both primary and secondary data were used for the study. Well structured questionnaire and interview schedule were used as approach to collect data from 80 respondents and six-stage modeling approach was used for assessing regional or landscape scale environmental impacts was adopted to analyze the impact of water irrigation, energy and environmental factors on agricultural productivity. The results showed that irrigation users are small-scale farmers, cultivating small hectare of land using simple farm tools, practices pump irrigation or calabash system. Regression analysis revealed that land, labour purchase inputs had a positive relationship with the output of the enterprises, while the linear programming analysis revealed that opportunities exist for increasing profit through resources re-organisation, budgetary analysis was used to show that irrigation farming is profitable.

Keywords: *Effect, Water, Irrigation, Energy, Environment, Agriculture, Productivity, Photovoltaic's, Hydropower, Biomass.*

1. INTRODUCTION

Irrigation agriculture is crucial to the economy, it is too important to be marginalized as it vital for world food security. However, irrigated agriculture often radically changes land use and is a major consumer of freshwater. Recognizing that the full potentials of Nigeria agriculture could not be realized without the development of her water resources for irrigation, governments in Nigeria have adopted various development policies (Palmer and Philip [1]).

Agriculture has an important impact on the environment. Pollutant sources from land use and poorly managed agriculture is causing widespread deterioration of ground water and surface water, while over irrigation results in water logging and the consequent accumulation of salts will lead to salinization of top soils and groundwater if not provided with proper drainage. Intensification of agriculture may lead to an increase in degradation of water quality and irrigated land because of an increasing intensity of inputs in agriculture (Aremun and Ogunwale, [2]).

The study revealed that water is scarce in the northern part of Nigeria; there will be an increase in use of brackish water and sewage effluents for irrigation. Recycling of water although may improve efficiency of use of water, it could also reduce both surface and groundwater quality. However, increased water withdrawals will lead to salt intrusion in estuaries and over-draught coastal aquifers. Maximum management of water quality is required to prevent these conditions and there is the need for proper monitoring of inflows and effluent water, in the same manner, recycling and reuse systems should take into account its cumulative effect overtime and its impact on the soil and water environment (Edwards, [3]).

The protection of water is not based on the prevention of pollution alone, but also to the recovery, conversation and the development of their self-purifying power. At this level, the restoration of rivers and lakes is an important task to be accomplished.

1.1 Problem Statement

A critical investigation of the performance of irrigation farming in Nigeria reveals that sub-sector are deficient in technical, financial, institution and human resources which subject farmers to persistent in the vicious cycles poverty, due to low income from low productivity and investment. This study examined the constraints encountered by the irrigation users and provided solution to increase resource utilization. Therefore, these questions are fundamental to this study:

1. What extents have the farmers thrive in irrigation farming using small-scale irrigation technologies?

2. How can urban and irrigation interest work together on how to meet future water needs of Nigeria

3. How can renewable energy technology such as biomass, hydropower, photovoltaic's, wind power and other technologies be made available for the benefits of farmers use

The main objective of this study therefore is to analyze the economic impact of water, irrigation, energy and environmental factors on agricultural productivity in Nigeria.

Specifically, the study:

1. Highlights the socio-economic characteristics of irrigation users,

2. Identifies the type(s) of crop enterprises engaged by the irrigation farmers in the study area

3. Determines how urban and irrigation interest can work together on how to meet future water needs for Nigeria

4. Determines how renewable energy, technologies such as biomass, hydropower, photovoltaic's, wind power and other technologies can be made available for the benefits of farmers use

1.2 Justification of the Study

The benefits of irrigation have resulted in lower food prices, provision of employment, more rapid agricultural and economic development. The spread of irrigation has been a key factor behind the near tripling of global grain production; therefore, there is need for a critical assessment of the performance of irrigation farming in the study area in order to achieve more of benefits of irrigation.

2. THEORETICAL FRAMEWORK

Farmers using irrigation have been identified as a key source of agricultural growth and development as opposed to previous irrigation development policies according to river Basin development, where irrigation schemes were designed and managed by government agencies for farmers. Irrigation has been long practiced in northern Nigeria where farmers have traditionally undertaken irrigation through the use of such technologies and methods as shadouf, buckets and calabash to produce high value agronomic and horticultural crops which are widely grown such as rice, sugar cane, cocoyam, leafy vegetables among others diverse cropping system. Fruits trees like citrus, mango and cashew, etc are planted on agricultural lands, this provides cash income as well as food crops to the farmers (Ohikere, [4]).

Worlf, [5]) observed that irrigation has made higher and more reliable yield possible as crops can be planted more than once in a year within the topics, apart from bigger and reliable yield as against yearly cultivation, which is often at the mercy of seasonal rainfall.

3. MATERIALS AND METHODS

The study was conducted in Patigi local government area of Kwara state, the population at 2006 census was 45,494 (22,712 males and 22,782 females. The major inhabitants are Nupe people while some are Yoruba and Hausa. They engage in planting of crops such as rice, sugar cane, groundnut, millet, melon etc. The climate is characterized by a distinct dry and wet season with annual rainfall varying from 1,200mm to 1,600mm. It is located approximately on the intersection of latitude 8°44' North and longitude 5°44'.

3.1 Sampling Procedure and Data Collection

A multi-stage random technique was used to select the respondent. Firstly, the study area was stratified into 8 cells. The second stage involved random selection of four cells out of the eight cells. The third stage involved random selection of two villages from each cell making a total of 8 villages. The last stage involved random selection of 10 respondents from each village making a total of 80 respondents. The sample size of the study was eighty irrigation users. The choice of this number was on the basis of the preliminary survey of the study area. The choice of random sampling techniques in the last stage was to give equal opportunity or chances to each farm family of being selected. Data for this study were collected from the selected farmers for the 2012/2013 irrigation season by the researcher and assisted by extension agents using well structured questionnaire and interview schedules. Data were collected between December, 2012 and January, 2013.

3.2 Measurement of Variables

There are two major variables in this study: they are the dependent variable and independent variable. The dependent variable is the total farm output, while independent variable includes land, labour, capital, irrigation water and socio-economic characteristics such as age, sex, marital status, educational level, households size etc. the resource constraints and irrigation water. The various levels of constraints were determined by what the "representative" farmers in the study area had. The representative farmers in the study area were taken to be the farmers who used the arithmetic mean of each of the resources. This view was supported by (Tsoho, [6]) who reported that "A representative farm can be used to depict a typical the farm in the sample" he further noted that although representative farms are often synthesis in the sense that none of them depict an actual farm, their component can be found on the majority of the farms they represent.

3.3 Data Analysis

Descriptive statistics such as tabulations, frequency distributions, mean (averages) were used to capture the first and second objectives. The third and fourth objectives were achieved by the use of multiple regression, logit model analysis and linear programming techniques of data analysis.

3.4 The Multiple Regression Model

The coefficient of multiple determinations (R^2) was obtained as a measure of goodness of fit. It is the percentage of total variation of the dependent variable (Y) explained by the variation in the independent variable (X_1-X_n). The equation with the higher R^2 value explain the variation better and was regarded a better fit. The regression coefficients for the various inputs used was tested using chi-square test to determine the inputs to which output is highly responsive, only those variables whose variable whose computed t-values are significant at 1 per cent, 5 per cent and 10 per cent was taken to have significant influence variation in output.

The Multiple Regression Equation and Explanation of Variables the Multiple Regression Equation and Explanation of Variables

$$Y = \alpha + \beta_1 X_1 + \beta_2 X_2 + \beta_3 X_3 + ... + \beta_n X_n + \varepsilon$$

where,

Y = (total farm output) dependent variable.

X_1= Age of Respondents

X_2= Sex (Male of Female)

X_3 = Marital Status

X_4 = Educational Level

X_5 = Household Size

X_6 = Land (Resource constraints in the study area)

X_7 = Labour (Resource constraints in the study area)

X_8 = Capital (the monetary value of machinery and tractors)

X_9 = Irrigation (method of supply water into dry land)

X_{10} = Water (water within wetlands, rivers, groundwater and ditches)

ε = error term

3.5 The Linear Programming Model

Linear programming is often helpful in decisions requiring a choice among a large number of alternatives. It has been employed in several studies to determine the optimal organization of peasant farming system in Nigeria (Ogunfowora, [7]).

The linear programming model employed for the study area was estimated as.

$$MaxZ = \sum (p_j q_i - c_i)$$

$$= \sum A_i X_{ji} - b$$

$$= X_i (j = 1 - M)$$

where,

Z = Return of owners labour and management (N/ha)

p_j = Price of jth crop per unit in naira (N)

q_j = Quantity of jth crop in kilogram

c_j = Total variable cost of labour and purchased inputs

A_j = Per unit requirement of jth activity carried out

M = The number of activities and it ranges from 1-4

i^{th} = Resource, ranges from 1-4h

b_i = The level of jth resources

b_1 = Average farm size (ha)

b_2 = Average labour available per farm in man-day/ha

b_3 = Average capital employed per farmer in man-day/ha

b_4 = Average water input in cm/ha

4. RESULTS AND DISCUSSION

4.1 Socio-economic Characteristics

Table 4.1 shows that over 81 percent of the respondents were males with females accounting for only 17.3 percent. This confirms the popular belief in the area that farming is a primary occupation for the male folks while the female folks are only to prepare food and maintain home for the males while working on their farms. It also confirms the religions belief that women in pedal are not to leave their homes for any outside activities.

Table 4.1: Socio-economic Characteristics of the Respondents

Characteristics	Frequency	Percentage
Gender Distribution of Respondents		
Male	66	81.5
Female	14	17.3
Total	**80**	**100.00**
Age Distribution of Respondents		
Age (group)		
25-40 years	11	8.4
41-50 years	32	39.3
51-60 years	33	40.8
60 Above	4	4.8
Total	**80**	**100.0**

Source: Field survey, 2013

Middle age group = 40-50 years; Mean age group = 40 years

Table 4.1 also shows the age distribution of respondents. Age is the length of past life or existence of a person. It is an important factor to be considered in determining the quality of labour employed and the labour force prevalent in any given enterprise. Age is particularly important considering the tedious nature of manual farming. In the study area, more than half of the respondents (80 per cent) were within the age groups of 25-60 years. Because of the tedious nature of manual farming which characterized the farming system in the area, only adults of working age (25-50years) could take into irrigation farming using pump, calabash and pipe watering irrigation methods. The age group also represents the most economically active age group.

The reason for low percentage of youth/young farmers (8.4 per cent) could be due to rural-urban migration and the quest for modern education urban centers.

Table 4.2: Educational Status of Respondents

Characteristics	Frequency	Percentage
Level of Education		
Primary	25	31.2
Secondary	22	26.7
Tertiary	11	13.4
No formal education	22	28.7
Total	**80**	**100.00**

Source: Field survey, 2013

Table 4.3: Marital Status of Respondents

Characteristics	Frequency	Percentage
Marital status		
Married	61	75.3
Divorced	8	9.9
Widowed	6	7.3
Single	2	2.5
Total	**80**	**100.00**

Source: Field survey, 2013

Table 4.4: Family Size of Respondents

Characteristics	Frequency	Percentage
House size		
1-5	28	34.7
6-10	40	49.1
11-15	8	10.1
>16	5	6.1
Total	**80**	**100.00**

Source: Field survey, 2013

Average Family size = 8; Standard Deviation= 4.8

Education; is a very important factor in development of any country's economy

The quality of skills of the farmers determines their level of locative abilities and show well informed of innovations and technology around him. Roger and Shoemaker [8] reported that education is not only an important determinant of adoption of innovations but also tools for successful implementation of innovation. Table 4.2 shows the educational status of the respondents in the study area, it shows that (28.7 per cent) of the respondents had no formal education. This is in accordance with the reports of (Tsoho, [6]). Agricultural production required marital status and it become an important factor especially when farm labour is in short supply. It is possible for married couple with large family size to have large supply of labour to work on the farm and this may increase the size of farm land cultivated. Table 4.3 also shows that (75.3) percentage or respondents (60.7 per cent) in the study area were married having average family size of 8 persons and hence some satisfied family labour for use on their farms.

4.1 Levels of Resource Use/impact of Input or Output

Factors that influenced output (Q) of the entire crop enterprises were regressed against the explanatory variables farm size in ha (X_1), labour input in mandays (x_2), value of purchased inputs in Naira(X_1) and the random error term (μ).The regression analysis result for each of the four functional forms examined in the production unction was linear, semi-log, exponential and double log of equations. From the regression estimate, the lead equation selected was Double- Logarithm (Cob-Douglas). This was based on fitness criteria, for example, apart from the fact that the it has a relatively higher R^2 value, it's also has two significant expected signs of regression coefficients both in terms of economic and agricultural logic, and is considered as the appropriate form to the estimated (lead) equation is presented as follows:

$$Q= -0.485+4.682 \times_1+0.572 \times_2+0.317 \times_3$$

$$= -0.485+(4.682)+(1.144)+(0.951)$$

$$= (-0.485)(13.815)(3.851)(2.089)$$

$$R^2 = 0.85$$

$$F = 94.433$$

From the equation, land and labour were significant ($P<0.05$). Also, land, labour and inputs have a positive relationship with output. Henceforth, more land, labour and inputs should be allocated to irrigation users in order to affect the much needed increase in output, ensure food security in the study area.

Table 4.5 shows the optimum results obtained through multiple regression, linear programme and logit regression model analysis for irrigation users (farmers) in the study area. It can be seen that out of the various crops combinations listed on the table above, cassava and maize entered the optimum. The Table 4.5 shows that 97.5 per cent of the respondents indicate that irrigation increase their crop yield, while 95.1 per cent claimed to have maximum food security as a result of irrigation methods practice in the study area. The difference in the variation of these impacts could be attributed to level of farm soil fertility. It was revealed from the table that 49.5 per cent of the respondents indicated that drought increase the stunted growth of their crop, this impact may be attributed to individual farmer level of understanding. It was revealed that 49.7 per cent of the respondents practiced ditch irrigation, 12.3 per cent used terraced irrigation, and 20 per cent used drip irrigation while 37 per cent, 6.2

Table 4.5: Crop Cultivated by the Respondents

Crop Cultivated	Frequency	Percentage
Maize	61	75.3
Sorghum	29	35.3
Rice	10	12.3
Cassava	75	93.8
Yam	59	72.8
Vegetable	23	28.4
Others	36	44.4
Total	**80**	**100.0**

Impact of Irrigation

Crop Cultivated	Frequency	Percentage
Decline the crop yield	8	9.9
Increase the crop yield	79	97.5
Stunted crop growth	2	2.5
Food shortage	7	8.6
Food secure	77	95.1
Food price increase	4	4.9
Food price decrease	40	49.4
Others	21	25.9
Total	**80**	**100.0**

Impact of Drought

Crop Cultivated	Frequency	Percentage
Food shortage	40	49.4
Decline in crop yield	61	75.3
Stunted crop growth	76	93.8
Decrease in land area cultivated	15	18.5
Poor quality of crops	34	42.0
Others	24	42.0
Total	**80**	**100.0**

Contd...

Table 4.5–Contd...

Irrigation Method

Crop Cultivated	Frequency	Percentage
Ditch irrigation	37	49.7
Terraced irrigation	10	12.3
Drip irrigation	16	20
Sprinkler irrigation	3	3.7
Rotary irrigation	5	6.2
Center irrigation	9	11.25
Total	**80**	**100.0**

Problem Encountered

Crop Cultivated	Frequency	Percentage
Maintenance problem	77	95.1
Water availability	40	49.1
Insufficient irrigation materials	21	25.1
Others	30	37.0
Total	**80**	**100.0**

Result of Logit Regression Analysis	Co-efficient	Standard Error	T-ratio
Age	37.86	9.647	3.925***
Sex	-39.65	24.41	-1.626*
Marital status	-1.166	24.41	-1.338
Household size	189.5	82.36	2.302**
Primary occupation (farming)	76.23	126.2	0.604
Total	**14.4**	**65.42**	**2.22****

Source: Field Survey Data, 2013

Log-likelihood = -65.08

Dependent variable = Irrigation user (Rep. by 1), while non irrigation user (Rep. by 0)

***: Significant at 1 per cent; **: Significant at 5 per cent; *: Significant at 10 per cent

and 11.25 per cent used sprinkle irrigation, rotary and center irrigation. The variation in the method used by the individual farmer depends on the cost and level of understanding of the irrigation users (farmer) in the study area as indicated by the farmers are maintenance problem 95.1 per cent, water availability 49.1 per cent, insufficient irrigation materials 25.1 per cent as it was revealed on the table with their percentages.

It was tested to confirm that irrigation and drought have both positive and negative significant effects on agricultural productivity. Also it was revealed on table 4.5 that age is statistically significant at 1 per cent with a coefficient value of 37.86 per cent with a positive sign, sex is significant at 10 per cent with a negative sign.

It further shows that household size is statistically significant at 5 per cent with a value of a 189.5 with a positive sign, while total land area are also significant at 5 per cent with a value of 14.4 per cent. Household size has the highest coefficient value which implies that it existed as the most important factor that greatly determined the output of the farmers in the study area.

5. CONCLUSIONS

The study examined the effect of water irrigation, energy and environment factor on agricultural productivity. The optimum irrigation farming computed showed that re organization, well planned and proper irrigation application would yield more income to farmers

6. ACKNOWLEDGEMENTS

I acknowledge the efforts of Mr. Ayegbusi Bamidele my supervisee who joined me on the study field to collect data for this study.

REFERENCES

1. Palmer, J.L and Philip, D.O. The economics of fertilizing wheat in two major Nigerian irrigation schemes in Maurya. Proceeding of the National Workshop held at IAR. Samam, Zaria. PP. 189-198. (1990)

2. Aremu J. A, Ogunwale, S. a (1994). Comparative Analysis of Small and large scale irrigation schemes in Northern Nigeria. Fact Finders Int. pp. 165-185. (1994).

3. Edwards, G. A/grove, T. C, Harwood, R.R, Pierce colfer, O. J. The role of Agro-Ecology and Integrated Farming System in Agriculture sustainability, Agric. Ecosyst. Environ. 46 (4):99-121. (1993).

4. Ohikere, J.Z and Ajogwu, Felix Ejeh. Impact of small scale irrigation technologies on rop production by Fadama users in Kogi State, Nigeria. Applied science Research 3 (2) 854-86. (2012).

5. Worlf, P. The problem of sustainability of irrigation system. Applied Geogr. 45(46) 55-62. (1995).

6. Tsoho, B.A. "Economics of Tomato-based cropping systems under small-scale irrigation in Sokoto State, Nigeria" Unpublished Msc. Thesis submitted to the Department of Agricultural Economics and Farm Management, University of Ilorin, Ilorin, Nigeria. pp. 87-95. (2005)

7. Ogunfowora, O. Capital and Credit market in Nigerian Agricultural Development Paper No 8. Department of Agricultural Economics and Extension, University of Ibadan. 5: 15-18. (1980).

8. Rogers, E.M. and Shoemaker, P.C. Diffusion of innovation. The free press of Glencoe, N.Y. PP. 367. (1971).

SECTION II
Renewable Energy Programmes

Chapter 5

Bio-fuels: An Alternative Energy Source towards a Sustainable and Eco-friendly Future: The Indian Scenario

Subhashree Basu

Research Assistant,
Centre for Science and Technology of the Non-Aligned and Other Developing Countries (NAM S&T Centre), New Delhi, India
E-mail: subhashreebasu@gmail.com

ABSTRACT

The world today is facing declining liquid fuel reserves with rapid increase in energy demands. As supply is dwindling and costs are rising, the nations are forced to utilize alternative energy sources. In order to achieve a secure and stable energy supply that does not cause environmental damage, exploration of renewable energy sources and development of promising appropriate technologies is a need of the hour. In this context, increasing concerns around energy security, depletion of fossil fuel, dependence on foreign oil and gas resources and negative consequences associated with climate change have prompted significant interest in the development of and investment on Sustainable Advanced Fuels with low carbon emission.

Bio-fuels from local feedstock can make a positive contribution to improving access to sustainable and affordable energy. Cultivation and harvesting of fuel crops can enhance agricultural productivity and local economic development directly as well as indirectly through crop by-products. Liquid bio-fuels like biodiesel and bioethanol can make way to the creation of an eco friendly transportation fuel that is relatively clean on combustion and provides farmers with substantial income. Sources such as *Jatropha*, algae, castor oil and other non-food feedstocks are a vast source of bio-fuels and a key to reducing our dependence on fossil fuel. The physical properties of biodiesel are similar to those of petroleum based fuels but it is a better alternative being a cleaner-burning fuel with significant reduction in the emissions of toxic air pollutants. Both the developed and developing countries are showing increasing trends in promoting the use of bio-fuels since this alternative fuel can help create local economic revitalization and global environmental benefits.

This paper reviews the rapidly evolving energy sector of India and assesses the current development of bio-fuels in the country with a view to identifying the production potential and constraints in its growth. It covers the range of feed stocks that are currently being considered for bio-fuels, and also discusses the agronomic and economic factors that enhance or limit their future growth. The paper also highlights some policy reforms and interventions that are being taken by the Indian Government which are necessary to the future efficiency and continued viability of this sector.

Keywords: Bio-fuels, Energy security, Renewable energy source, Alternative fuel.

1. INTRODUCTION

Ever-increasing demand for energy across the globe has led to a vast reduction in the non renewable energy sources. The biggest problem is the growing population and depletion of fossil fuel. Our dependence on fossil fuels has caused carbon dioxide (CO_2) enrichment of the atmosphere and is the primary contributor to global warming. In order to realize a stable energy alternative that will meet world demand while mitigating climate change, it is necessary to develop renewable clean fuels. The world's oil production is expected to decline in between one and ten decades (Crookes, 2006). Other non-renewable sources of energy exist, such as coal and uranium; however, these sources are limited and will also inevitably decline in availability. As a result of this impending energy crisis, both governments and private industry are examining alternative sources of energy. Though, a number of other alternatives like ocean water power, geothermal energy, wind energy and solar energy are being explored, bioenergy is viewed as a strong source of energy in the coming years. Bioenergy is the energy generated through bio-fuels that are produced from renewable sources of plant origin. Ironically, most renewable energy initiatives are focused on electricity generation, while the majority of world energy consumption, about two thirds, is derived from liquid fuels (Hankamer *et al.*, 2007). Today the world is moving towards a bio-based economy which is defined as an economy that uses renewable bio-resources, efficient bioprocesses and eco-industrial clusters to produce sustainable bio-products, jobs and income.

Countries moving towards a bio economy will provide new opportunities for industry and farmers and will reduce greenhouse gas emissions. Bio-based products include fuels, energy, chemicals, lubricants, plastics, paper construction materials and advanced composites.Numerous major companies have increased their involvement in the development of bio-based products. The United States (US) National Research Council predicts 50 per cent of US fuels and 90 per cent of organic chemicals will come from renewable resources by the turn of the century.

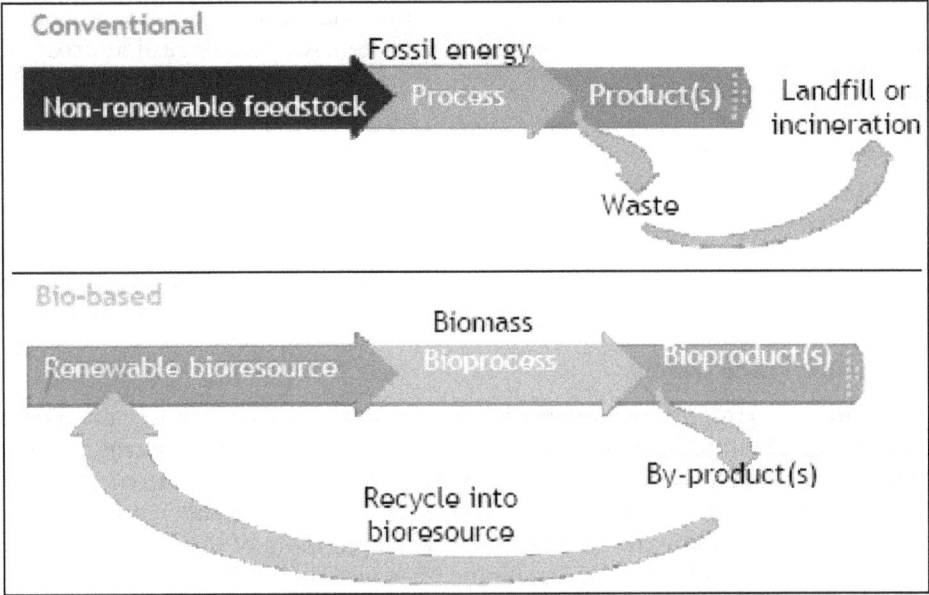

Figure 5.1: Comparing the Structures of Conventional vs. Bio-based Economies

The increasing energy demands and decreasing sources of petroleum and concerns about climate change have created a need for focused bio-fuel research across many countries (Perlack *et al.*, 2005; Himmel *et al.*, 2007). The need for renewable sources of portable, liquid fuel is starting to receive greater attention, and much of this attention has been focused on biomass-derived liquid fuels, or bio-fuels (Schneider, 2006; Haag, 2006).

This paper examines the feasibility of biodiesel and bioethanol as a potential replacement for petroleum-based liquid fuels. In particular, the use of *jatropha* seeds and other non-food feedstocks for fuel production is investigated, in terms of its productivity, practicality, and innovative potential to create a cost competitive, environmentally friendly, and renewable source of liquid fuel. Since India is a major agriculture based country hence it is important to see the prospective of using Bio-fuels in India. In this context, this paper has addressed the evolving bio-fuels sector of India, and the growth of bio-fuels as alternatives to fossil-based transportation fuels. The paper has outlined the key issues that confront the future development of this new sector and has identified the production potential and constraints of bio-fuels development.

2. BIO-FUELS

BIO-FUELS are liquids or gaseous fuels produced from biomass resources and used in place of, or in addition to, diesel, petrol or other fossil fuels for transport, stationary, portable and other applications. Bio-fuel is a fuel that contains energy from geologically recent carbon fixation. Bio-fuels have increased in popularity because of rising oil prices and the need for energy security. Either in liquid form such as fuel ethanol or biodiesel or in gaseous form such as biogas or hydrogen, bio-fuels are the transportation fuels derived from the renewable biological sources.

Importance of Bio-fuels

1. Energy Independence

Considering that oil priced at $ 60 per barrel has had a disproportionate impact on the poorest countries, majority of which are net importers and few of which import all of their oil, it has become crucial to achieve greater energy independence through the development of bio-fuels and, now on a near daily basis, a bio-fuels programme is being launched somewhere in the developing world.

2. Smaller Trade Deficit

Rather than importing other countries' ancient natural resources, we could be using our own living resources to power our development and enhance our economies. Instead of looking to the Mid-east for oil, the world could look to the tropics for bio-fuels. Producing more bio-fuels will save foreign exchange and reduce energy expenditures and allow developing countries to put more of their resources into health, education and other services for their neediest citizens.

3. Economic Growth

Bio-fuels create new markets for agricultural products and stimulate rural development because bio-fuels are generated from crops; they hold enormous potential for farmers, who derive their incomes from agriculture. Today, many of these farmers are too small to compete in the global market, especially with the playing field tilted against them through trade distorting agricultural subsidies. They are mostly subsistence farmers who, in a good year, produce enough to feed their families, and in a bad year, grow even poorer or starve. But bio-fuels have enormous potential to change this situation for the better. At the community level, farmers that produce dedicated energy crops can grow their incomes and grow their own supply of affordable and reliable energy. At the national level, producing more bio-fuels will generate new industries, new technologies, new jobs and new markets.

4. Cleaner Air

Bio-fuels burn more cleanly than gasoline and diesel. Using bio-fuels means producing fewer emissions of carbon monoxide, particulates, and toxic chemicals that cause smog, aggravate respiratory and heart disease, and contribute to thousands of premature deaths each year.

5. Less Global Warming

Bio-fuels contain carbon that was taken out of the atmosphere by plants and trees as they grew. The Fossil fuels are adding huge amounts of stored carbon dioxide (CO_2) to the atmosphere, where it traps the Earth's heat like a heavy blanket and causes the world to warm. Studies show that bio-diesel reduces CO_2 emissions to a considerable extent and in some cases all most nearly to zero. Biodiesel fuel burns up to 75 per cent cleaner than diesel fuel made from fossil fuels. Biodiesel substantially reduces unburned hydrocarbons, carbon monoxide and particulate matter in exhaust fumes. Sulphur dioxide emissions are 100 per cent eliminated.

Types and Sources of Bio-fuels

The bio-fuels are broadly classified as first generation bio-fuels and second generation bio-fuels. *First generation bio-fuels* are sourced from the biomass consisting of sugars, starch, vegetable oils, animal starch, or other biodegradable products from agriculture, industry, forestry and households, using conventional technologies. Cereals like maize, sweet sorghum and sugar crops like sugarcane, sugar beet, etc. can be easily fermented to produce ethanol, which can be used either as a motor fuel in pure form or as a blending component in gasoline. Some oilseeds both edible (rapeseed, soybean, sunflower, oil palm) and non-edible (*Jatropha*, pongamia, neem, etc.) can be converted into a liquid fuel which can be blended with the conventional diesel fuel or burnt as pure biodiesel. However, the first generation bio-fuels have some limitations. The major one is that they cannot be produced beyond a threshold level without threatening food security. They are also not cost-competitive with the existing fossil fuels. The combating of these problems led to the search for *second generation bio-fuels* which are more sustainable, affordable and environment-friendly. These comprise ligno-cellulosic materials, including vegetative grasses, trees, waste products from crops and wood processing facilities, and municipal solid wastes; these can be converted to fuel alcohol.

Recently, technologies for converting algae into ethanol have also been developed. But the process is more complex relative to the processing of sugars and grains. Techniques are being developed to convert the cellulosic crops and crop wastes to ethanol. Similarly, biodiesel produced from organic waste materials, including used cooking oils and biogas produced from animal manure and organic household wastes can be categorized under second generation bio-fuels. Technologies are also under development for commercial production of bio-fuels like biohydrogen, biomethanol, butanol and isobutanol etc. produced from different types of biomass feedstock. Bio-fuels offer a number of environmental, social and economic advantages, apart from being a renewable alternative for fossil fuels.

Bioethanol

Bioethanol is an alcohol made by fermentation, mostly from carbohydrates produced in sugar or starch crops such as corn or sugarcane. Cellulosic biomass, derived from non-food sources, such as trees and grasses, is also being developed as a feedstock for ethanol production. Ethanol can be used as a fuel for vehicles in its pure form, but it is usually used as a gasoline additive to increase octane and improve vehicle emissions.

Ethanol Production

Production of ethanol requires two steps: fermentation and distillation. However, not all feedstocks such as starch based (corn, cassava, wheat, etc.) and cellulose based (wood materials, agriculture waste, black liquor from pulp and paper, bagasse, forestry waste) contain simple sugars. In both cases hydrolysis (addition of water, *e.g.* enzymes) is needed to convert starch/cellulose into simple sugars (glucose) followed by fermentation of glucose to ethanol. In terms of cost, fermentation of starch and ligno-cellulose to ethanol is more due to additional complex and costly investments for converting them into ethanol.

Global ethanol production is monopolized by two major producers, the USA, which uses corn starch, and Brazil, which uses sugarcane sucrose. In both cases, this production is based on food sources. The proportion of food sources for the production of bio-fuels reaches a critical point when an impasse is reached between the production of raw material for fuel ethanol or for food. This impasse represents a bottleneck in the maintenance and expansion of the bio-fuels market. One of the short-term alternatives would be to use these plants solely for food and use only their lignocellulosic materials for the production of ethanol. This would help mitigate environmental pollution and minimize the use of food sources for ethanol production.

Biodiesel

Biodiesel is made from vegetable oils and animal fats. Biodiesel can be used as a fuel for vehicles in its pure form, but it is usually used as a diesel additive to reduce levels of particulates, carbon monoxide, and hydrocarbons from diesel-powered vehicles. Biodiesel is produced from oils or fats using trans-esterification and is the most common bio-fuel in Europe.

Biodiesel refers to a non-petroleum-based diesel fuel consisting of short chain alkyl (methyl or ethyl) esters, made by Trans-esterfication of vegetable oil or animal fat (tallow), which can be used (alone, or blended with conventional petrodiesel) in unmodified diesel-engine vehicles. Biodiesel is distinguished from the straight vegetable oil (SVO) (sometimes referred to as "waste vegetable oil", "WVO", "used vegetable oil", "UVO", "pure plant oil", "PPO") used (alone, or blended) as fuels in some converted diesel vehicles. "Biodiesel" is standardized as mono alkyl ester.

Biodiesel is an environmentally safe, low polluting fuel for most diesel internal combustion and turbine engines. Biodiesel is an eco-friendly, alternative diesel fuel prepared from domestic renewable resources that is vegetables oils like oilseed, rapeseed and soya bean (edible and non-edible oil) and animal fats. These natural oils and fats are made up mainly of triglycerides. These triglycerides when reacted chemically with lower alcohols in presence of catalyst result in fatty acid esters. These esters show striking similarity to petroleum derived diesel and are called "Bio-diesel". Bio-diesel is the most valuable form of renewable energy that can be used directly in any existing, unmodified diesel engine.

Chemically, bio-diesel is made up of mono-alkyl esters of long chain fatty acids. It is produced by transesterification (Figure 5.2) in which, oil or fat reacts with methanol in the presence of alkaline catalyst (*e.g.* sodium hydroxide). The reactions occurs stepwise triglycerides first converted into diglycerides, then to monoglycerides and finally to glycerol (Chisti, 2007).

Advantages of using Biodiesel

1. *Easy to use*: Biodiesel can be used in existing engines, vehicles and infrastructure with practically no changes. Biodiesel can be pumped, stored

$$
\begin{array}{ll}
\underset{|}{CH_2}-O-\overset{\overset{O}{\|}}{C}-R_1 & R_1-\overset{\overset{O}{\|}}{C}-O-CH_3 \quad CH_2-OH \\
\underset{|}{CH}-O-\overset{\overset{O}{\|}}{C}-R_2 + 3CH_3OH \longrightarrow & R_2-\overset{\overset{O}{\|}}{C}-O-CH_3 + CH_2-OH \\
CH_2-O-\overset{\overset{O}{\|}}{C}-R_3 & R_3-\overset{\overset{O}{\|}}{C}-O-CH_3 \quad CH_2-OH
\end{array}
$$

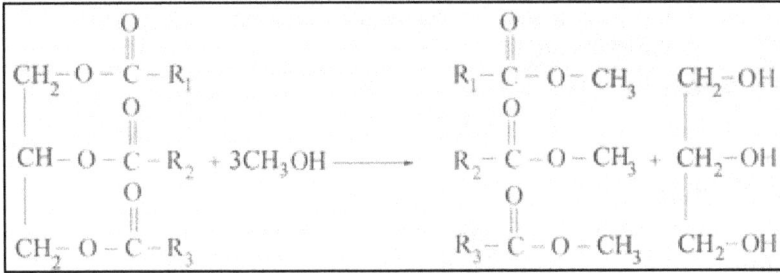

Figure 5.2: Transesterification of Oil to Biodiesel
R is hydrocarbon group

and burned just like petroleum diesel fuel, and can be used pure, or in blends with petroleum diesel fuel in any proportion.

2. *Power and Performance*: The degree to which fuel provides proper lubrication is its lubricity. Low lubricity petroleum diesel fuel can cause premature failure of injection system components and decreased performance. Biodiesel provides excellent lubricity to the fuel injection system.

3. *Emissions and Greenhouse Gas Reduction*: Biodiesel provides significantly reduced emissions of carbon monoxide, particulate matter, unburned hydrocarbons, and sulfates compared to petroleum diesel fuel. Additionally, biodiesel reduces emissions of carcinogenic compounds by as much as 85 per cent compared with petrodiesel. When blended with petroleum diesel fuel, these emissions reductions are generally directly proportional to the amount of biodiesel in the blend.

Greenhouse Gases: Carbon dioxide produced from biodiesel combustion does not contribute to new emissions of CO_2 as it is part of the carbon cycle. Being a closed carbon cycle it results in 80 per cent reduction in CO_2.

Smog-Forming Pollutants (Ozone precursors): Compared to diesel, biodiesel effects on engine exhaust are in the following table that follows: B100 refers to neat biodiesel; B20 refers to 20 per cent biodiesel blended with Diesel:

- HC   36.73% reduction with B100   7.35% reduction with B20
- CO   46.23% reduction with B100   9.25% reduction with B20
- PM   68.07% reduction with B100   13.61% reduction with B20
- SOx(3)   100% reduction with B100   20% reduction with B20

The above data on HC, CO, PM, and SOx is based on data reported by Dr. Groboski in 1998 paper "Combustion of Fat and Vegetable Oil Derived Fuels in Diesel Engines.

4. *Economic Development*: The resources that are used to produce biodiesel are locally available. The in-house production of biodiesel provides host of economic benefits for the local communities.

- Creation of more employment, Jobs created for feedstock farming and/ or collection

- Skilled jobs created for biodiesel production and distribution (Engineers, Technicians etc.)
- Tax Benefits
- Generation of Income for the local feedstock producers and refiners.

Global Scenario

Global bio-fuel production grew from 16 billion litres in 2000, to more than 100 billion litres in 2010. This bio-fuel provides around 3 per cent of the world's fuel for transport. In Brazil, bio-fuel provides 23 per cent of all transport fuel, compared with 4 per cent in the United States and 3 per cent in the European Union (IEA 2011). Currently, around 82 per cent of the global production of liquid bio-fuels is in the form of ethanol. The two largest ethanol producing countries, Brazil and the USA, accounted for almost 87 per cent of the total production, the rest being contributed by China, Canada, France, India, Russia, South Africa, UK, etc. (Figure 5.3).The highest per hectare yield in ethanol production has been realized by Brazil which has set up an economically competitive national bio-fuel sector largely based on sugarcane. In Brazil, the yield of bioethanol is around 5,476 litres/ha (74.5 litres/tonne of sugarcane), which is higher than that of any other country. Brazil has developed a cost-effective technology of converting sugarcane juice directly into ethanol, in contrast to that of many other countries including India where ethanol is produced mostly from molasses a by product of sugar industry. In comparison, the yield of maize-based ethanol in USA and China is much lower, it is around 3,751 litres/ha, and 1,995 litres/ha, respectively (Naylor *et al.,* 2007). In China, wheat, cassava and sweet sorghum are

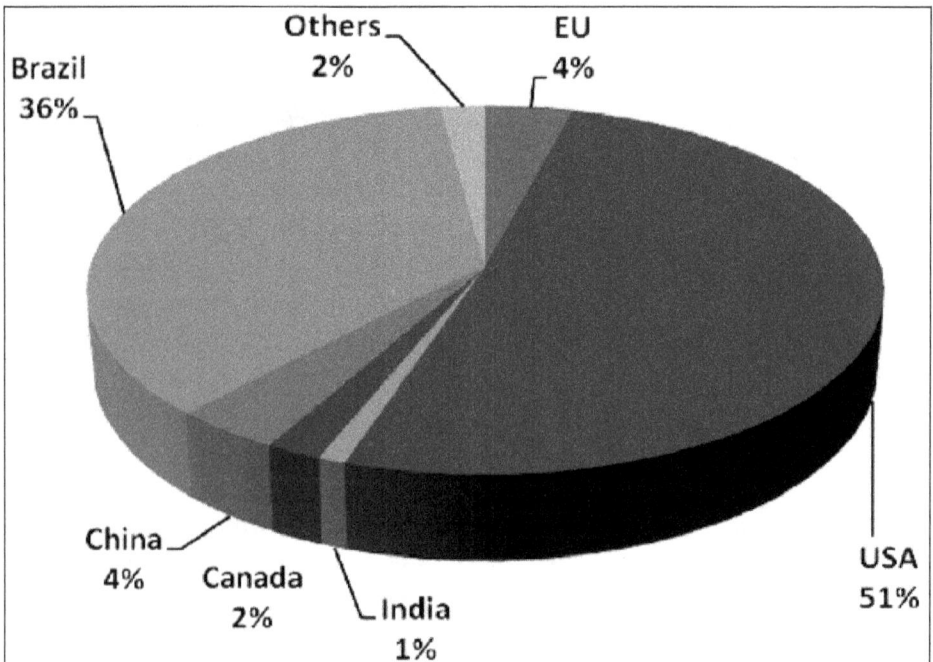

Figure 5.3: Ethanol Production by Country: 2010

used besides corn for ethanol production. European Union (EU), another major ethanol producer, uses cereals like wheat, corn, barley and sugarbeet for production of bio-ethanol. Blending rates differ substantially across the countries. While USA mandates 3 per cent blending of ethanol with petrol, Brazil is following a very high ratio of 25 per cent blending. China and Indonesia have set a target of 10 per cent blending, whereas in EU the blending specification stands at 5.75 per cent in the year 2010.

Biodiesel production that accounted for a smaller proportion of liquid bio-fuels, increased from 0.01 million tonnes in 1991 to 21.0 million tonnes by 2010. European Union is the major producer of biodiesel (above 50 per cent), with a significantly smaller contribution coming from the USA (17 per cent). Other major biodiesel producers include China, India, Indonesia and Malaysia (Figure 5.4). In EU, 80 per cent of the biodiesel is produced from rapeseed oil, the rest being animal fats and other used cooking oils. Oil palm is the major source of diesel production in Malaysia and Indonesia, whereas both USA and Brazil are using soybean to produce biodiesel (Table 5.1). In India, biodiesel production is only at the nascent stage, with about 95 million litres being produced from *jatropha* and pongamia oil. The bio-fuel production programme in most of the countries is being supported through subsidies. In Organisation for Economic Co-operation and Development (OECD) countries, both production and consumption of bio-fuels are being subsidized, mainly under the banner of energy security and climate change mitigation. It is estimated that, on per litre basis, the support ranges between US $ 0.20 and US $ 1.00. Among the major producers, only the Brazilian sugarcane ethanol appears to be competitive without subsidies (FAO, 2008). The International Energy Agency has a goal for bio-fuels to

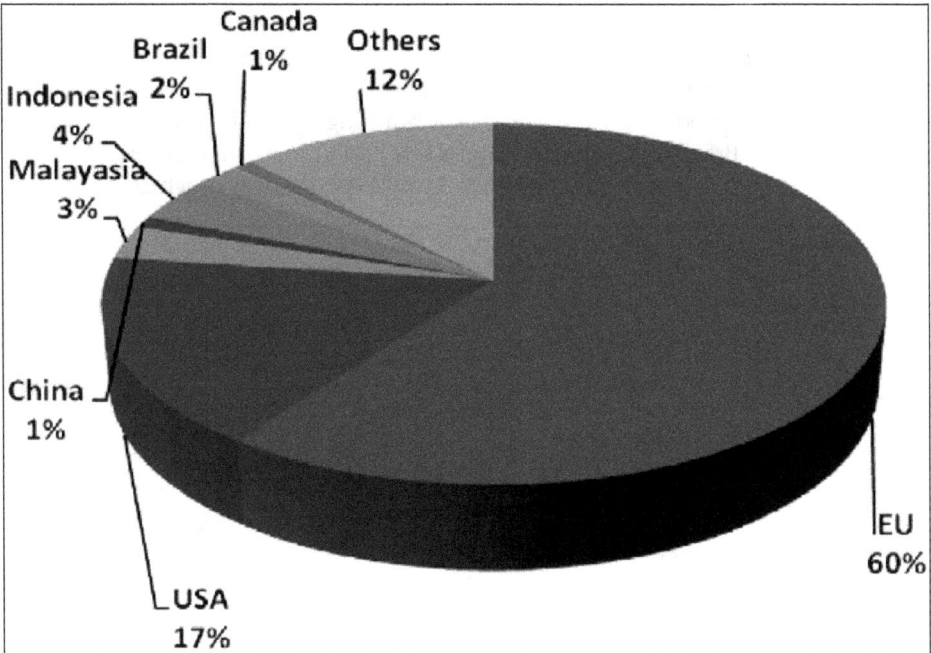

Figure 5.4: Biodiesel Production by Country: 2010

meet more than a quarter of world demand for transportation fuels by 2050 to reduce dependence on petroleum and coal.

Table 5.1: Bio-fuel Feedstocks and Blending Targets in some Selected Countries

Country	Feedstock		Production Forecast 2010 (million litres)		Blending Targets (Per cent)	
	Ethanol	Biodiesel	Ethanol	Biodiesel	Ethanol	Biodiesel
USA	Corn	Soybean	46017	2707	3	1
Brazil	Sugarcane	Rapeseed, castor seed	28950	2162	25	2
EU	Wheat, corn, barley, sugar beet	Rapeseed, sunflower, soybean	6465	9888	5.75	5.75
Canada	Corn, wheat	Vegetable oils	1572	360	5	2
China	Corn, wheat, cassava, sweet sorghum	Palm oil, *Jatropha*	2083	–	10	5
India	Sugarcane molasses, sweet sorghum	*Jatropha*, pongomia	1550	95	5	5
Indonesia	Sugarcane, cassava	Palm oil, *Jatropha*	425	348	10	10
Malaysia	None	Palm oil	–	647	–	5

Source: F.O. Licht (2009); FAPRI (2010)

The global market for biodiesel is poised for explosive growth in the next ten years. Although Europe currently represents 80 per cent of global biodiesel consumption and production, the U.S. is now ramping up production at a faster rate than Europe, and Brazil is expected to surpass U.S. and European biodiesel production by the year 2015. It is possible that Biodiesel could represent as much as 20 per cent

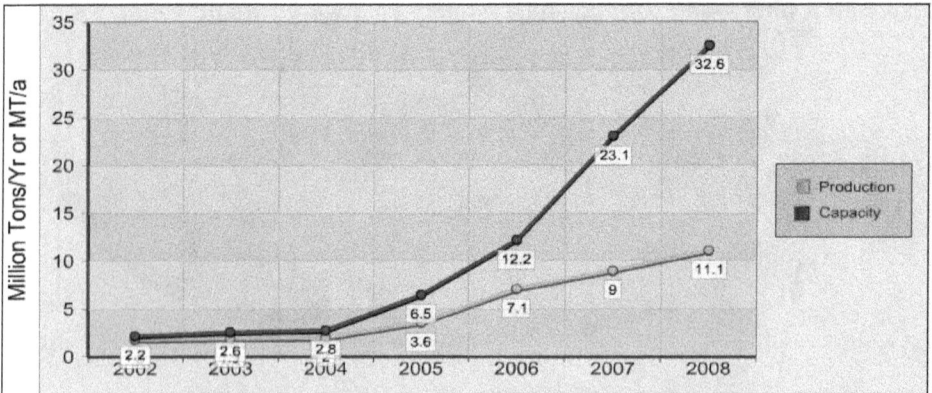

Figure 5.5: World Biodiesel Production and Capacity
Source: **Biodiesel 2020: A Global Market Survey**

of all on-road diesel used in Brazil, Europe, China and India by the year 2020 with the pursuit of second generation, non-food feedstocks. Biodiesel demand and over-capacity in Europe, the US and Asia is driving investment in the global trade of alternative feedstocks.

Bio-fuels in India: Potentials, Policies and Emerging Paradigms

The Indian Scenario

India is one of the fastest growing economies in the world. The Development Objectives focus on economic growth, equity and human well being. Energy is a critical input for socio-economic development. The energy strategy of a country aims at efficiency and security and to provide access which being environment friendly and achievement of an optimum mix of primary resources for energy generation. Fossil fuels will continue to play a dominant role in the energy scenario in our country in the next few decades. However, conventional or fossil fuel resources are limited, non-renewable, polluting and, therefore, need to be used prudently. On the other hand, renewable energy resources are indigenous, non-polluting and virtually inexhaustible. India is endowed with abundant renewable energy resources. Therefore, their use should be encouraged in every possible way.

India is the world's fifth largest primary energy consumer (as per international energy annual) and fourth largest petroleum consumer after United States, China and Japan. The growing population, increasing per capita income, infrastructural development and rapid socio-economic development have spurt an increase in energy consumption across all the major sectors of the Indian economy. Given the limited domestic energy resources, escalating crude oil prices, and growth in domestic consumption of petroleum products, India's oil import bill has inflated considerably. In the recent past, starting in Indian fiscal year (IFY) 2009/10, imports of gasoline and petroleum products has outgrown total domestic consumption by more than 14 percent. While India's domestic energy base is substantial, the country continues to rely on imports for a considerable amount of its energy use, consequently escalating India's oil import expenditure to over $135 billion in IFY 2011/12, up 22 percent over the previous year (Figure 5.6). Concurrently, petroleum consumption in India has also grown in tandem to 148 million tons. (Petroleum Planning and Analysis Cell,Government of India (GOI)). The only alternative is a switch to an alternative fuel source, such as bio-fuels. However, bio-fuels sector in India is still at its nascent stage and is evolving.

The growth of bio-fuels around the globe is spurred largely by energy security and environmental concerns and a wide range of market mechanisms, incentives and subsidies have been put in place to facilitate their growth. Developing countries, apart from these considerations, also view bio-fuels as a potential means to stimulate rural development and create employment opportunities. The Indian approach to bio-fuels, in particular, is somewhat different to the current international approaches which could lead to conflict with food security. It is based solely on non-food feedstocks to be raised on degraded or wastelands that are not suited to agriculture, thus avoiding a possible conflict of fuel vs. food security.

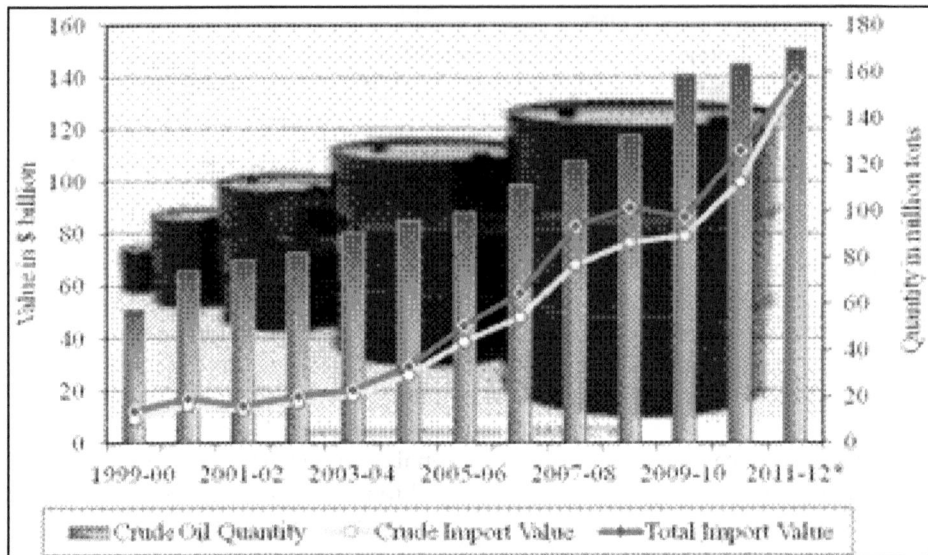

Figure 5.6: India: Import of Crude Oil and Value of Petroleum
Source: Petroleum Planning and Analysis Cell, Government of India (GOI)*: estimate

Indian Bio-fuel Initiatives

The Bio-fuel Policies and Potential in India

The increased move towards Bio-fuels is spurred as an urgent need for renewable and eco-friendly source of energy which could assist in enhancing the self-sufficiency in energy and minimizing dependence of a nation on imported fossil fuels. Towards this endeavour, the Government of India has initiated several programmes to augment production and use of bio-fuels during the past decades or so. The National Bio-fuel Mission, launched in 2003, is the frontrunner of such initiatives, with Ethanol Blended Petrol Programme (EBPP) and Biodiesel Blending Programme (BDBP) as its integral components. In these programmes, specified, time-bound targets have been laid for blending of 5 per cent, 10 per cent and 20 per cent bio-fuels with petrol and diesel in a phased manner so as to catalyze the transition from a completely fossil fuel based transport system to a partially bio-fuel-driven system. The feedstocks identified are molasses, for the production of ethanol and tree-borne nonedible oilseed crops like *jatropha* and pongamia for the production of biodiesel. India has some special advantages in taking up plantation of tree-borne oilseeds for the production of biodiesel as the country has vast under-utilized or unutilized land, either, fallow, barren, degraded or understocked, most of which are in the drought-prone areas or are otherwise unsuitable for growing of food crops. To give momentum to the bio-fuel sector, the Government of India has announced various support measures like minimum purchase price for fuel ethanol as well as biodiesel, minimum support price for *jatropha* seeds, etc. The 'National Policy on Bio-fuels', released in 2009, foresees bio-fuels as a potential means to stimulate rural development and generate employment opportunities, as well as aspires to reap environmental and economic benefits arising out of their large-scale use. It is categorically mentioned in the Policy that the bio-fuels programme is to be carried out based solely on the non-food feedstocks that are raised on the degraded or wastelands not suitable for agriculture, thus avoiding a possible conflict between food security and fuel security.

Mission, Vision and Goals of the Bio-fuel Policies*

☆ National Bio-fuel Policy aims at mainstreaming of bio-fuels and, therefore, envisions a central role for it in the energy and transportation sectors of the country in coming years

☆ Strengthen India's energy security by encouraging use of renewable energy resources to supplement transport fuels

☆ The Goal of the Policy is to ensure that a minimum level of bio-fuels become readily available in the market to meet the demand at any given time. An indicative 20-percent target for blending of bio-fuel for both biodiesel and bioethanol, by 2017 is proposed

☆ Meet the energy needs of India's vast rural population, stimulating rural development and creating employment opportunities

☆ Address global concerns about net reduction of carbon emissions through use of environmentally friendly bio-fuels

☆ Derive bio-fuel from inedible feedstock grown on degraded soils or wastelands unsuited to food or feed production, thus avoiding a possible conflict of fuel- versus food security

☆ Facilitate optimal development and utilization of indigenous biomass feedstock for production of bio-fuels. The policy also envisages development of next-generation, more efficient bio-fuel conversion technologies based on new feedstock

Ongoing Projects

In concurrence with the larger bio-fuel promotion programme undertaken by the central government at the national level, various state governments are also pursuing policies and programmes in support of bio-fuels production and use in their respective constituencies. Some states have even drafted their own policy documents and vision statements which though distinct, are in line with the broad spirit of the National Bio-fuel Policy. For instance, the state of Chhattisgarh has started a comprehensive programme on *jatropha* production in wastelands and forest lands. For coordinating and monitoring various biodiesel-related activities in the state, Chattisgarh Bio-fuel Development Board has been created. The Board encourages the local resource-poor farmers for taking up large-scale cultivation of *jatropha* in leased-in lands with active participation of local governing bodies and NGOs. Some private entrepreneurs have also joined hands by setting up processing facilities in the *jatropha*-growing areas. Similar state organizations have been constituted in the states of Rajasthan and

---

* *Source*: National Policy on Bio-fuels. Ministry of New and Renewable Energy, Govt. of India, pp. 18; 2009

Uttarakhand also and various models of *jatropha* cultivation are being practised. Several central and state sponsored programmes like MGNREGS (Mahatma Gandhi National Rural Employment Guarantee Scheme), NWDPRA (National Watershed Development Programme in Rainfed Areas), etc. are being linked to bio-fuel crops cultivation in various ways. The Governments of Tamil Nadu, Andhra Pradesh, Odisha and Karnataka are also encouraging cultivation of bio-fuel crops by framing favourable bio-fuel policies. Most of these states have announced minimum support prices for feedstock seeds, though these are not uniform across the states. Various types of tax concessions, subsidy schemes, etc. are being extended to bio-fuel cultivators and processing industries in these states for faster adoption of the policy prescriptions.

In India, biodiesel is produced mainly from non-edible oilseed crops like *jatropha* and pongamia, edible oil waste and animal fats. Currently, *jatropha*, the major feedstock for biodiesel, occupies only around 0.5 million hectares of wastelands across the country, of which 65-70 per cent are new plantations of under three years. The study has shown that the farmers do not have any incentive to divert their fertile lands for *jatropha* cultivation at present. The Report of Planning Commission on Development of Bio-fuels (GoI, 2003) has earmarked an estimated area of 13.4 million hectares of marginal/wastelands that are suited to growing *jatropha* and this can cater to large-scale plantings so as to meet the blending targets fixed by the Government of India. Similarly, the Department of Land Resources under the Ministry of Rural Development, has estimated that around 25 million hectares of fallow land is available in the country that can be diverted for growing of feedstock crops, including *jatropha*. Given these facts, there is no apprehension about a threat to food security from commercial bio-fuel blending programmes in India as opposed to the case with other major bio-fuel producing countries.

Bioethanol

India is globally one of the largest producers of sugarcane and ethanol made from sugarcane molasses. For producing ethanol, India has about 330 distilleries with the annual production capacity of over 4.0 billion litres. In the year 2010, the country produced nearly 1.43 billion litres of ethanol, of which an estimated 50 million litres of ethanol were blended with petrol. Ethanol is primarily produced by the fermentation of molasses, and it is estimated that, from one tonne of sugarcane, 85-100 kg of sugar (8.5–10 per cent) and 40 kg (4 per cent) of molasses can be obtained. The recovery of ethanol from molasses is 22-25 per cent as per Indian standards. Presently, about 70-80 per cent of sugar cane produced in the country is utilized for the production of sugar and the remaining 20-30 per cent for alternative sweeteners (jaggery and khandsari) and seeds. Thus, only molasses produced during sugar production are available for ethanol production. Due to the cyclical nature of sugarcane and thus sugar production in India, sugarcane farmers and the processing industry experience periodic market gluts/deficit of sugarcane, sugar and molasses, impacting their prices and farm income.

Biodiesel

Unlike other countries, India is not using vegetable oils derived from rapeseed and mustard, soybean or oil palm for the production of biodiesel. It is because, India is not self-sufficient in edible oils production and depends upon imports of palm oil and other vegetable oils in large quantities to meet the domestic demand. Every year around 1.2 million tonnes of tree borne non-edible seed oils are produced in the country (GoI, 2008). In India, biodiesel is produced mostly from the non-edible oils extracted from the seeds of plants like *jatropha* and pongamia. Many of the strong proponents of the use of *jatropha* for producing oil-based bio-fuel, claim a number of potential benefits which have been widely cited in the literature (GOI, 2003; UNCTAD, 2006; ADB, 2011). These are:

☆ The oil produced is non-edible, which helps to avoid food versus- fuel tradeoffs

☆ It has high oil content (40 per cent) and relatively low gestation period (2-3 years)

☆ It can survive in areas of low rainfall and low fertility

☆ It requires less inputs and minimum care for cultivation

☆ The cake obtained after oil extraction can be used as organic manure, for preparation of herbal insecticides and biogas

☆ *Jatropha* has the ability to assist in the build-up of soil carbon

☆ *Jatropha* is an animal deterrent crop and hence not browsed by grazing animals

The actual realization of these benefits requires a number of economic and agronomic conditions, on the ground, to be met – and remains somewhat elusive for some regions that have tried to scale up *jatropha* cultivation for large scale production. However, the current biodiesel industry in India is still at a nascent stage although the Government has ambitious plans to expand this sector.

Challenges Faced

Even though various measures have been undertaken to provide support, progress in the bio-fuels sector, its development has been rather slow in India. A major issue is that India's ethanol programme depends largely on the economic viability of 'molasses-ethanol conversion'. Recent studies have noted that the present economics of molasses-based ethanol production do not favour commercial blending of ethanol in petrol. The current cost of production and the pricing structure of ethanol favours its diversion towards potable purposes than for blending with gasoline. Because of this reason, the oil marketing companies have so far not been able to achieve mandatory blending targets fixed by the government.

Another concern is regarding the long-term sustainability of molasses based ethanol blending programme. Studies have indicated that if the government is targeting to bring into effect 10 per cent blending by the year 2016–17, as planned in the National Bio-fuel Policy, production of approximately 736.5 million tonnes of sugarcane with area coverage of 10.5 million ha would be required. This means, both production and area under sugarcane will have to be more than doubled to achieve 10 per cent blending target. Given the current trends in yield and area growth, achieving the 20 per cent blending target appears still more unlikely without significant imports of ethanol. Moreover, it would be highly unsustainable to extend the sugarcane area beyond a certain limit, given the fact that sugarcane is a highly water-intensive crop with water requirement of 20,000–30,000 m³/ha/crop. An alternative to improve the efficiency of ethanol recovery is direct conversion of sugarcane juice ethanol; but since this option would be at the cost of reduction in sugar production, and therefore is again an impracticable proposition. Thus, it is high time to think about diversifying to alternative sources of sugar and ethanol such as sweet sorghum, tropical sugarbeet, etc. which are both resource saving and sustainable.

The uncertainty regarding the conveyance of ownership of community and government wastelands for cultivation of *jatropha* and other tree-borne oilseeds is another major issue that require prompt care. Lack of good quality planting materials, low confidence among farmers to expand cultivation due to uncertainty regarding profitability, lack of an integrated approach for cultivation, and so forth, are impeding the expansion of area under *jatropha*. The production potential of *jatropha* and other bio-fuel crops is unclear as wide fluctuations in yield levels depending on the agro-climatic conditions, management, etc., have been noticed. Lack of sufficient seed processing infrastructure is presently a major constraint holding back the development of biodiesel sector in India. In most of the *jatropha*- growing areas, modern processing plants have not come up in sufficient numbers so far. This is because of two major reasons: first, the government intends bringing private participation to build this capacity, but the private players visualize potential risks in investing in this area because of uncertainty regarding the supply of sufficient feedstock and market demand for biodiesel. Second, the unavailability of processing capacity is making the farmers to down-scale their production and this poses a threat to even the existing processing plants. The cost of production of biodiesel increases substantially if the units are run under low economies of scale. The problem worsens with an increase in the price of seeds due to the involvement of middlemen and higher transportation costs when the seeds are sourced from distant places. A proper value chain for marketing and processing of *jatropha* seeds, distribution of biodiesel, financing services, institutional and governance arrangements are yet to be developed.

Other issues like pricing and taxing of bio-fuels also need immediate attention for the development of the existing bio-fuel sector in the country.

Need of Hour: Regulatory, Fiscal and Policy Framework

An important driving force for bio-fuels development is a conducive regulatory fiscal and policy framework. This is required to ensure that sustainable development benefits of increased bio-fuels development are properly reflected in holistic approach being proposed; including markets, distribution, production and economics.

Some of the policy issues that need elaboration include the following:

☆ Awareness and information programme

☆ Articulation on out-grower policy to enhance sustainable feedstock production and supply

☆ Articulation on land-use issues

☆ Articulation on policy of replacement of lead as an octane enhancer with ethanol

☆ Harmonisation of standards on biodiesel and ethanol

☆ Establishing organisational structures to transport or deliver bio-fuels product

☆ Assessment of socio-environmental impacts

☆ Tax incentives on technology and bio-fuels

Interventions and Enabling Mechanisms

GOI is planning to take up various initiatives to expand the Bio-fuel Sector:

Plantations

1. Plantations of trees bearing non-edible oilseeds to be taken up on Government/community wasteland, degraded or fallow land in forest and non-forest areas. Contract farming on private wasteland may also be taken up through the Minimum Support Price mechanism proposed in the Policy

2. Quality seedlings of non-edible oilseeds to be raised in the nurseries of certified institutions/organizations identified by the States for distribution to the growers and cultivators

Processing Support

Oil Marketing Companies in 20 states and 4 Union Territories in India are responsible for blending 5 per cent ethanol with gasoline. The sugar industry has been permitted to produce and process ethanol from sugarcane juice to augment production to meet blending requirements.

Output Support

Output support for the purchase of bio-fuels is vital. The National Bio-fuels Policy by GOI proposes a Minimum Support Price mechanism for *Jatropha* whose seed is used to produce biodiesel. In the case of sugarcane, the existing statutory minimum price provides effective protection to growers.

Distribution Support

Oil Marketing Companies have been responsible for the storage, distribution and marketing of bio-fuels in India. India's bio-fuel policy exempts the bio-fuel sector from central taxes and duties. While biodiesel is exempt from excise duty, bioethanol enjoys a concessional excise duty of 16 per cent. Customs and excise duty concessions are also provided on plant and machinery for the production of biodiesel and bioethanol. These policies promote the bio-fuel sector (Shinoj P, 2011).

Financial Incentives and Subsidies

Financial institutions like the National Bank for Agriculture and Rural Development (NABARD), Indian Renewable Energy Development Agency (IREDA) and Small Industries Development Bank of India (SIDBI) have refinancing provisions to set up biodiesel plantations, oil expelling/extraction units, and infrastructure for storage and distribution (Reddy BVS, 2008). In India, most inputs like fertilizer, pesticide and electricity to irrigate crops are subsidized.

Consumption Support

Under Section 52 of India's Motor Vehicles Act, an existing vehicle's engine can be modified to use bio-fuels. Hence, engine manufacturers need to suitably modify engines to ensure compatibility with bio-fuels. Demand for such vehicles and bio-fuels may be stimulated by providing road tax exemption and reduced registration fee for vehicles running on blended fuels.

Research and Development and Demonstration

The policy's major thrust is innovation, Research and Development (R&D) and demonstration. Grants should be provided to academic institutions, research organizations, specialized centers and industry for promising R&D and demonstration projects for Public Private Partnership (PPP).

1. Research and Development must focus on plantations, bio-fuel processing and production technologies, as well as on maximizing efficiencies of different end-use applications and utilization of by-products.

2. Intensive R&D work would be undertaken in the following areas (As per the Bio-fuel Policy initiative):

   (a) Bio-fuel feed-stock production based on sustainable biomass with active involvement of local communities through non-edible oilseed bearing plantations on wastelands to include inter-alia production and development of quality planting materials and high sugar containing varieties of sugarcane, sweet sorghum, sugar beet, cassava, etc.

   (b) Advanced conversion technologies for first generation bio-fuels and emerging technologies for second generation bio-fuels including conversion of ligno-cellulosic materials to ethanol such as crop residues, forest wastes and algae, biomass-to-liquid (BTL) fuels, bio-refineries, etc.

   (c) Technologies for end-use applications, including modification and development of engines for the transportation sector based on a large scale centralized approach, and for stationary applications for motive power and electricity production based on a decentralized approach.

   (d) Utilisation of by-products of bio-diesel and bio-ethanol production processes such as oil cake, glycerin, bagasse, etc.

3. For R&D and demonstration projects, grants would be provided to academic institutions, research organizations, specialized centers and industry. Strengthening of existing R&D centers and setting up of specialized centers in high technology areas will also be considered. Linkages would be established between the organizations/agencies undertaking technology development and the user organizations. Participation by industry in R&D and technology development will be encouraged with increased investment by industry with a view to achieve global competitiveness.

3. With regard to Research and Development in the area of bio-fuels, a Subcommittee under the Bio-fuel Steering Committee proposed in this Policy comprising Department of Bio-Technology, Ministry of Agriculture, Ministry of New and Renewable Energy and Ministry of Rural Development would be constituted, led by Department of Bio-Technology and coordinated by the Ministry of New and Renewable Energy.

Role of States

The role and active participation of the States is crucial in the planning and implementation of bio-fuel programmes. Certain States have already set up agencies that are suitably empowered and funded to act as nodal agency for development and promotion of bio-fuels in their States. Other concerned agencies, panchayati raj institutions, forestry departments, universities, research institutions etc. would also need to be associated in these efforts. While a few States have announced policies for bio-fuel development, other States would also need to announce suitable policies in a time-bound manner in line with the broad contours and provisions of the National Policy.

Institutional Mechanisms

1. Under the Allocation of Business Rules, the Ministry of New and Renewable Energy has been given the responsibility of Policy and overall Coordination concerning bio-fuels. Apart from this, the Ministry has also been given the responsibility to undertake R&D on various applications of bio-fuels. Responsibilities have also been allocated to other Ministries *viz.* Ministry of Environment and Forests, Ministry of Petroleum and Natural Gas, Ministry of Rural Development and Ministry of Science and Technology to deal with different aspects of bio-fuel development and promotion in the country.

2. In view of a multiplicity of departments and agencies, it is imperative to provide High-level co-ordination and policy guidance/review on different aspects of bio-fuel development, promotion and utilization. For this purpose, a National Bio-fuel Coordination Committee (NBCC) headed by the Prime Minister was set up to provide high-level coordination, effective end-to-end implementation and monitoring of bio-fuel programmes.

CONCLUSIONS

☆ Bio-fuels offer enormous potential as an alternative energy source. However time bound, coordinated efforts will be essential to exploit this potential optimally

☆ By using Bio-fuels, India has a ray of hope in providing National Energy security and can contribute for reducing the import bill of oil for the country

☆ The development of bio-fuels sector could result in increased cultivation of the feedstock crops like *jatropha* (*Jatropha curcas*), *pongamia* (*Pongamia pinnata*), etc. especially in marginal and forest lands and could provide higher income and employment opportunities for economically and socially backward communities that cultivate these crops

☆ To transfer the bio-fuel sector into an economically-viable venture in India, biotechnological innovations are essential

☆ An up-to-date technology policy is central to bring in efficiency in production which is also cost effective so that the industry would survive on its own without any subsidy or support

☆ Improved agronomic practices for sugarcane cultivation, along with research on suitable alternatives would help provide this sector the required impetus

☆ The focus on research needs to be sustained to explore the feasibility of environment-friendly and economically-sustainable feedstocks

☆ Research thrust to develop technologies for commercial production of next generation bio-fuels (*e.g.* biobutanol) from crop wastes should also go hand in hand

☆ Efforts should also be directed towards promoting community participation and entrepreneurship through higher participation of local institutions in the process of bio-fuel development

☆ Above all, it is important to ensure that the national policy on bio-fuels is comprehensive and is based on the pillars of economic viability, technological feasibility, environmental sustainability and market friendliness. Such a policy should also ensure energy security without compromising with food security of the nation. The policy framework to promote the bio-fuel sector in India is promising. But government's initiatives have not translated into results. Introspection into the possible causes and loopholes in the system needs to be done. The National Bio-fuel Policy should be regularly updated after incorporating latest trends and techniques as well as economic and legal aspects

REFERENCES

1. ADB (Asian Development Bank) (2011). Cross-Sectoral Implications of Bio-fuel Production and Use in India, Report submitted to Department of Economic Affairs, Ministry of Finance and Government of India, New Delhi.

2. Chisti, (2007). "Biodiesel from microalgae"; Biotechnology Advances Vol 25 pp 294–306

3. Crookes, (2006). "Comparative Bio-Fuel Performance in Internal Combustion Engines," Biomass and Bioenergy, Vol. 30, pp. 461-468.

4. FAO (2008). Bio-fuels: Prospects, risks and opportunities, The State of Food and Agriculture, Rome.

5. FAO (Food and Agriculture Organization) (2008). The state of food and agriculture. Rome, Italy. pp 128.

6. GoI (Government of India) (2003). Report of the Committee on Development of Bio-fuels, Planning Commission, New Delhi.

7. GoI (Government of India) (2008). Agricultural Statistics at Glance. Ministry of Agriculture, New Delhi.

8. GOI (Government of India). (2009). National Policy on Bio-fuels. New Delhi, India: Ministry of New and Renewable Energy. 18 pp.

9. Haag, A. L., "Algae Bloom Again," Nature, Vol. 447, May 31, 2007, pp. 520-521.

10. Hankamer, B.; Lehr, F.; Rupprecht, J.; Mussgnug, J. H.; Posten, C. and O. Kruse, (2007). "Photosynthetic biomass and H_2 Production by Green Algae: From Bioengineering to Bioreactor Scale-Up," Physiologia Plantarum, Vol. 131, pp. 10-21.

11. Himmel M. E. *et al.* (2007). Biomass recalcitrance: engineering plants and enzymes for bio-fuels production. *Science* 315, 804–807. *(doi:10.1126/science.1137016)*

12. IEA International Energy Agency (2011). "Technology Roadmap, Bio-fuels for Transport".

13. Kazi Mostafijur Rahman, Mohammad Mashud, Md. Roknuzzaman and Asadullah Al Galib; Biodiesel from *Jatropha* Oil as an Alternative Fuel for Diesel Engine; International of Mechanical and Mechatronics Engineering IJMME-IJENS Vol10. No 3

14. Naylor, R., Liska, A.J., Burke, M.B., Falcon, W.P., Gaskell, J.C., Rozelle, S.D. and Cassman, K.G. (2007). The ripple effect: Bio-fuels, food security, and the environment. Environment, 49(9): 31–43.

15. Perlack *et al.* (2005). Biomass as feedstock for a bioenergy and bioproducts industry: the technical feasibility of a billion-ton annual supply, A joint study sponsored by the U.S. Departments of Energy and Agriculture. DOE/GO-102005-2135, ORNL/TM-2005/66, April.

16. Planning Commission (2003). Report of the Committee on Development of Bio-fuels. Government of India, 10. New Delhi

17. Reddy BVS, Ramesh S, Ashok Kumar, A, Wani SP, Ortiz R, Ceballos H and Sreedevi TK. (2008). Bio-fuel crops research for energy security and rural development in developing countries. Bioenergy Research 1:248–258.

18. Schneider, D., "Grow your Own?: Would the Wide Spread Adoption of Biomass-Derived Transportation Fuels Really Help the Environment," American Scientist, Vol. 94, pp. 408, 409.

19. Shinoj P, Raju SS and Joshi PK. (2011). India's bio-fuel production programme: Need for prioritizing alternative options. Indian Journal of Agricultural Sciences 81(5):391-397.

20. UNCTAD (United Nations Conference on Trade and Development), (2006). An Assessment of the Bio-fuel Industry in India, Report prepared for UNCTAD Geneva.

21. www.mnre.gov.in/

Chapter 6

Portable Solar Anaerobic Digestion Pilot Plant for Biogas Production

Kifah Al-Imarah*, Mudher Hani, Taha Majed, Ismaeel Mohammed and Weleed Moufak

Ministry of Science and Technology,
Renewable Energy Directorate, Iraq
*E-mail: abdkefah@yahoo.com ; moderhani@yahoo.com

ABSTRACT

The research aims to adopt a strategy for producing biogas as a cheap and clean source of renewable energy into rural areas from hybrid anaerobic digestion after appropriate treatment of an organic waste, be burned directly for cooking, lighting or heating. It can also be used in an internal combustion engine for producing electricity.

Small- scale biogas digester (2.5 m^3, China origin) was loaded with cow manure and farmyard wastes then mixed with water to maintain a total solid content 10 per cent. The produced biogas was pumped from desulfurization filter to PVC storage bag through biogas pump device powered by charger solar panel (12 watts). The results showed that the rate of biogas is about (0.8 m ^{3}/m ^{3} digester volume). The biogas stove consumes (0.45 m^3/h) gas while the rice cooker consumes gas at a rate (0.14 m^3/h). The biogas lamp consumes gas at rate (0.07 m^3/h) and the electric generator consumption is (0.84 m^3/h).

By trapping and utilizing the methane, a greenhouse gas (GHG) impact is avoided. Further, the pathogens existing in manure are eliminated by the heat generated in the biodigestion process and the resulting material provides a valuable, nutrient-rich fertilizer.

Keywords: Solar charger, Desulfurization filter, Biogas pump and Biogas Generators

1. INTRODUCTION

Energy is one of the most challenging needs of humanity, and is highest on the list of priorities and requisites for human welfare. It has become a global priority to develop sustainable, efficient and inexpensive energy to meet demands. One of the major issues with large-scale public adoption of any alternative energy is its cost.

Alternative large-scale anaerobic digestion is a commercially proven technology and is widely used for recycling and treating wet organic waste and waste waters. It is a type of fermentation that converts organic material into biogas, which mainly consists of methane (approximately 60 per cent) and carbon dioxide (approximately 40 per cent) and is comparable to landfill gas Biomass (Conversion Technologies, ESMAP, 2005).

Anaerobic digestion is the natural biological process which stabilizes organic waste in the absence of air and transforms it into bio-fertilizer and biogas. It is a reliable technology for the treatment of wet, organic waste. Organic waste from various sources is biochemically degraded in highly controlled, oxygen-free conditions circumstances resulting in the production of biogas which can be used to produce both electricity and heat. Almost any organic material can be processed with anaerobic digestion. The exception to this is woody wastes that are largely unaffected by digestion as most anaerobic microorganisms are unable to degrade lignin (Beatrix, 2012).

Some of the wastes might be difficult to digest as a sole substrate, but they can be biochemically degraded in combination with manure or sewage sludge. The combined digestion of different wastes is called co-digestion (Salman, 2014).

Anaerobic digestion of farm-based biodegradable solids has been widely studied in the literature within full scale biomass-to-energy conversion (Balasubramanian, 1992; Chynoweth, 1987; Cantrell, 2008; Ward, 2008). As evidence, in European Union countries a total biogas generation equal to 36.5 billion mC^3 is anticipated as a target for 2020. The number of farm-based medium and large-scale biogas generation plants (MLBGP) in China has been increased from approximately 1500 units in 2004 to around 4700 units by the beginning of 2010 (Zhang, 2009). While China's target for 2020 is to generate 2 billion m^3 out of total 44 billion m^3 biogas from agricultural sources (GTZ, 2010).

Several technologies of anaerobic digestion including anaerobic contact process, completely mixed reactor, Upflow anaerobic sludge blanket (UASB), Upflow anaerobic hybrid blanket, anaerobic filter and plug-flow reactor are popular in China for handling livestock wastes and crop residues (He, 2010). However, the majority of the Chinese on-farm biogas plants visited by the authors, have been working with digester influents containing below 8 per cent total solids (TS) content whilst Chinese farmers often separated the larger solid constituents e.g. feces and straw from agricultural waste streams in advance and sold them as a fertilizer neglecting their significant potential of methane production as well as their potential pollution load in the terms of organic compounds, pathogens, odors and parasites.

This research describes the attempt for producing biogas from small-scale pilot plant composed of an anaerobic fermentation tank, a gas storage bag, a gas pump, a solar charger, as well as gas pipeline and appliances. The unit is easy and quick to install. It can be installed by the Users themselves within two hours and convenient to transport, because its total weight is only 80 kg. It is suitable to use food waste, grass, leaves and straw as fermentation material. High gas production rate, the gas production rate can reach $0.8m^3$.

2. MATERIALS AND METHODS

2.1 Feedstock Preparation

The materials used for this experiment were cow dung and farmyard waste. The cow dung was sun dried for a period of 20 days to preserve its microbial population. There are basic conditions needed for biogas production that can provide a strict anaerobic environment is the key to successful biogas production. The methane bacteria needed for anaerobic fermentation comes from anaerobic activated sludge (also called strain). The sludge from running biogas plants or cow dung is also inoculums. The methane bacteria ferment biomass is at the temperature between (35-45), and the higher the temperature the more active the methane bacteria, and the higher the biogas production. The pH value of the prepared liquid in the biogas plant is controlled between (6.8 and 7.5). The carbon-nitrogen ratio of the fermentation materials meet (20-30:1), the anaerobic bacteria will be active; therefore, the biogas production will be fast.

2.2 Construction Details of Solar Portable Biogas Pilot Plant

The configuration of pilot-scale ($2.5m^3$) family anaerobic digestion system has been schematically illustrated in (Figure 6.1).

The digester was designed in the form of a cylindrical chamber with (1.40m) height and (1.97m) length. The digester was equipped with inlet, outlet, and biogas tube at different levels and polar angles to each other. The reactor chamber divide into two compartments, (0.76m) the digester's height of upper tank and (0.65m) the height of the bottom tank.

The digester tank should be put on the site with enough sun shine. The site should be leveled with no objects on it. It is better to cover the leveled site with a 30-50mm thick layer of fine soil or sand. The installation of the outlet pipe by turning the lower tank up down and putting the pipe jointer on the outlet and then put the mixed AB glue in the gaps between the outlet and the jointer (it takes about one hour to dry the glue).

The installation of upper tank and lower tank by turning the lower tank, let the mouth up and clean up the sealing groove, then put the rubber sealing rope in the sealing groove. The length of the sealing rope depending on the length of sealing groove. Cut the two ends of the sealing rope into wedges of 45" and put the two ends together and wrap them up tightly by adhesive tape. Fix the sealing rope in the sealing groove with adhesive tape. Put the upper tank on the lower tank and fix them with stainless screws one.

In order to install of the gas conducting pipe, paint a layer of glue on the inner wall of the gas pipe jointer, and then insert the straight pipe in the pipe jointer. Paint

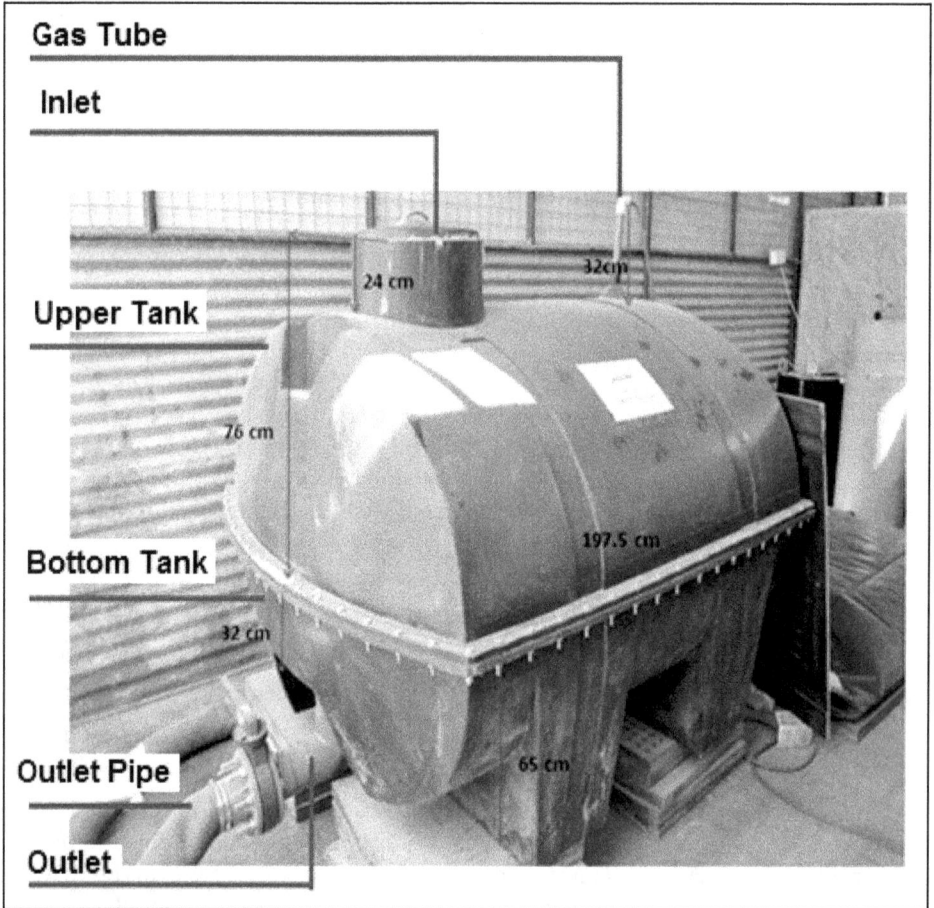

Figure 6.1: Portable Anaerobic Digester Pilot-Scale Unit

a layer of glue on the outside wall of the other end of the straight pipe in the elbow with the brass reducer which insert in the soft gas pipe.

The installation of the solar charger through electric wire which the length of the wire can be adjusted according to users' needs but should be no longer than (30 meters). The red color wire is positive pole. The gas pump can be run by either normal (220 V) power or the solar battery run gas pump, the positive pole of the solar battery should be connected to the positive pole of the plug of the gas pump. The installation of the remaining parts as shown in (Figure 6.2).

The first batch feeding and initial operation of the anaerobic digestion pilot plant, China origin, for the production of biogas from cow manure and farmyard wastes in a 2.5 m^3 digester made of fiberglass reinforced plastic to protect it from rust and corrosion. The gas was pump to PVC storage bag with (thickness 0.6mm), size (2 m^3) and dimension (2.8 x 2.8 meters). It is Easy and quick to install and can be installed by the users themselves within two hours. It is convenient to transport,

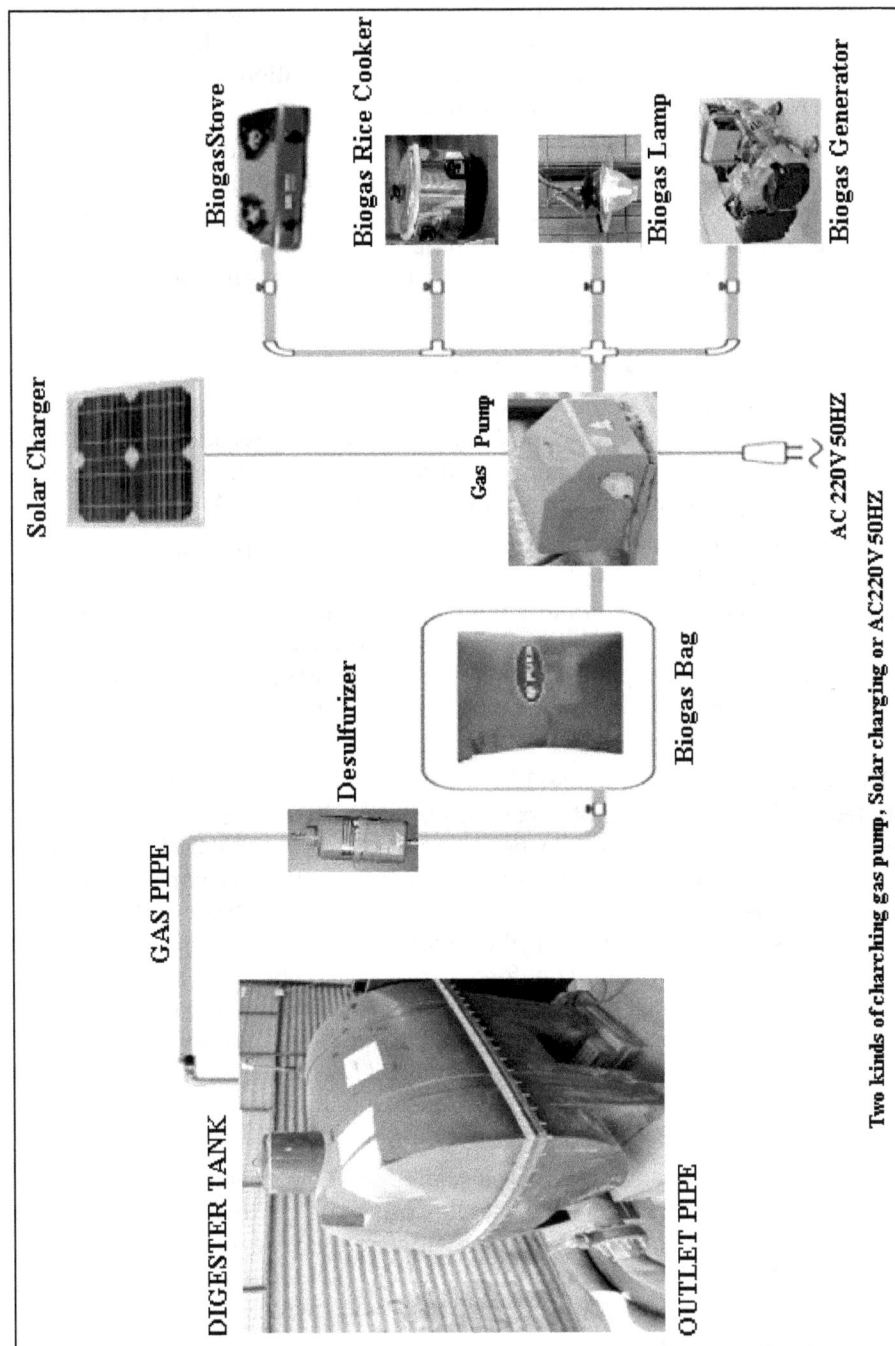

Two kinds of charching gas pump, Solar charging or AC220V 50HZ

Figure 6.2: The Installation of Solar Portable Anaerobic Digestion Pilot-Scale Unit

because its total weight is only 80 kg to the places where are wastes available, especially in rural areas. In addition, ease of cleaning and maintenance process. This system can be used above ground to prevent contamination of ground water especially in areas with the high level of water. Wide range of fermentation material: It is suitable to use food waste, grass, leaves and straw as fermentation material. High gas production rate: the gas production rate can reach $0.8m^3/m^3$.

The system contain filter to remove hydrogen sulfide gas and (12 W power) solar charger to biogas pump (10 W power and 26 L/min pumping rate) to gas stove double burner with consumption rate about $0.45m^3/h$ (for one burner) and 2.5L biogas rice cooker with gas consumption about $(0.14m^3/h)$. Biogas lamp with electronic fire maker which consume $(0.07 m^3/h)$ gas and an electric generator (600W) capacity with about $(0.84 m^3/h)$ gas consumption.

3. RESULTS AND DISCUSSION

China may be famous for the ability to use organic wastes for biogas production since the end of the 19th century and later it supported mass adoption of biogas mainly on household level. The company has developed a series new generation hydraulic biogas plants, and obtained a number of patents and Awards.

Selecting the Chinese design as the most appropriate type at the moment for small-scale, backyard livestock-raising families is dictated by the following reasons:(a) it is less costly, volume-for -volume, compared with other existing designs; (b) it is more durable and entails less maintenance for it has fewer metal components and moving parts; and (c) it is capable of delivering a higher gas pressure, thus making it possible to utilize the gas for cooking and lighting purposes, as well high quality organic fertilizers.

The configuration of pilot-scale $(2.5m^3)$ family AD system has been schematically illustrated in (Figure 6.1). Some of its basic features are circular in shape, with a fixed-dome top and saucer-shaped bottom. Being circular, it has the smallest lateral surface area with the pressure and load stresses even distributed, thus compact and solid. Additionally, it is economical in construction materials (up to 2) per cent less than a quadrangle or square structure having the same volume). No moving parts and metal components, except the structural reinforcements. An all concrete structure, hence, it is durable, almost a life-long investment. It is easier to maintain for it has no mechanical moving parts. No welding job is necessary. Input materials flaw easily into the digester, thus simplifying operation. Capable of generating higher gas pressure (on the average 10 times higher than floating cover type) is due to a unique gas storage mechanism which does not need a floating tank as other designs do.

The gas production rate of this digester can reach $(1.5m^3)$ per day. Therefore, the users can cook all their meals on what just they produce from their kitchen garbage. The system came with a sulfur removal cartridge to upgrade the biogas and a gas lamp, a typical pressure kerosene lamp (the kind that use a gauze mantel) with a battery electric starter, so that we don't need to use electricity to light the kitchen.

After the closure of the system, in normal cases is generated gas after 3-10 days. Attribute the lack of gas flaring in the beginning to lower methane content in the digester. If there is no flammable gas, should be launched every gas from the gas holder and re- assembled and this process is repeated until the flammable gas. In the absence of flammable gas for a long time or decreasing gas production quickly or even stop production of gas in full and turning color of the solution to the yellow color attribute the cause to acidic and there are several ways that can be used to modify the different levels of acidification adjustment. If the value of the function pH = 6, the acidity of the system automatically increases gradually becomes a natural gas production in this case gas production needs a longer period of time, so put the most amount of prefixes inoculums to the digester may reduce the time required to digest. If the value of the function pH > 6 should withdraw some of the liquid and add more prefixes inoculums to the digester and add some lime or ash in order to adjust the pH value above 6. If the value is set to acidic pH = 7 and still there is no production of methane gas, means that the liquid in the digester contains chemicals work on the killing of methane bacteria. In this case digester should be cleaned and re fill it with raw materials and prefixes inoculums together. In a normal operation, it is possible to increase the concentration of nutrients between (8-10 per cent) or re- use the liquid output of the digester for a second time.

4. CONCLUSIONS

Anaerobic digestion is a waste conversion process - it is not a waste "disposal" process. In fact there will be little reduction in volume of waste but the treated waste will be less smelly, contain less pathogen and be better as a plant fertilizer than untreated waste. You will also have the benefit of renewable energy released during the pollution reduction process.

The work in the digester distinguishes quickly install and ease of transport. It is suitable for use in the fermentation of organic materials such as food waste, grass, leaves, straw and animal waste with a high rate of gas production and can be up (0.8 m^3/m^3). The rate of removal of volatile solids VS was more than 85 per cent and the period of digestion ranged between (10-20) days with more than 75 per cent of the methane content in the biogas. This product need concentration research to develop some problems obtained, related to the design. The system should provide with an Insinkerator Food Waste Grinder (feedstock preparation device or compost companion) to make the organic waste that comes from your kitchen homogeneous liquid and easy to digest.

There are a number of safety issues to be considered when working with a biogas system. Methane (CH_4), form explosive mixtures in air, the lower explosive limit being 5 per cent methane and the upper limit 15 per cent methane. Biogas mixtures containing more than 50 per cent methane are combustible, while lower percentages may support, or fuel, combustion. With this in mind no naked flames should be used in the vicinity of a digester and electrical equipment must be suitable quality, normally "explosion proof". Other sources of sparks are any iron or steel tools or other items, power tools, normal electrical switches, and static electricity. If conducting a flammability test take a small sample well digester, or incorporate a flame trap in the supply line, which must be suitable length (minimum 20 m).

5. ACKNOWLEDGMENTS

We extend our thanks and appreciation to deputy general manager of Renewable Energy Directorate Mr. Diya Hussain and donors, especially Dr. T. H. Culhane to support this pilot- scale project, which will contribute to improving the standard of living to whom lives in poor and rural areas through the use of environmental resources in a sustainable manner.

REFERENCES

1. Biomass Conversion Technologies, ESMAP, 2005. Sustainable Bioenergy Development in UEMOA Member Countries.

2. Beatrix Rózsáné Szûcs, Miklós Simon and György Füleky, Co-Digestion of Organic Waste and Sewage Sludge by Dry Batch Anaerobic Treatment, Management of Organic Waste, http: www.intechopen.com. Published online 01, February, 2012.

3. Salman Zafar, Biogas feedstock in middle east. Report on the EcoMENA, Echoing Sustainability (accessed 14 January 2014)

4. Balasubramanian, P. R. and Katuri Bai, R., 1992. Evaluation of nutrient recovery after anaerobic digestion of cattle dung in a family size biogas plant. Biomass and Bioenergy, **3 (5)**:377-380.

5. Chynoweth, D. P. and Isaacson, R., 1987. Anaerobic Digestion of Biomass. Elsevier Applied Science, London, UK.

6. Cantrell, K. B., Ducey, T., Ro, K. S. and Hunt, P. G., 2008. Livestock waste-to-bioenergy generation opportunities. Bioresource Technology, **99**:7941-7953.

7. Ward, A. J., Hobbs, P. J., Holliman, P. J. and Jones, D. J., 2008. Optimization of anaerobic digestion for agricultural resources. Bioresource Technology, **99**:7928-7940.

8. Zhang, X., Wang, R., Huo, M. and Martinot, E., 2009. A study of the role played by renewable energies in China's sustainable energy supply, Energy, Doi:10.1016/j.energy.2009.05.030.

9. GTZ. The Sino-German project on optimization of efficient biomass utilization. Proceedings of Biomass Energy Utilization Seminar, Leipzig, Germany. (accessed 26 Apr. 2010)

10. He, P. J., 2010. Anaerobic Digestion: An intriguing long history in China. Waste Management, **30**: 549-550.

Chapter 7

Sustainable Energy for All: Transforming Commitments into Action-DOST MIMAROPA Region of the Philippines

Bernardo Tandang Caringal

Provincial Science and Technology Director-Marinduque
DOST MIMAROPA, 2nd Floor Imelda Building,
Bicutan, Taguig City, Philippines
E-mail: dostmar@yahoo.com

ABSTRACT

Adversities concerning natonal calamities such as typhoons which usually lead to problems in economic development particularly power interruptions became the focus of the Government of the Republic of the Philippines for the development of programs and projects advocating renewable energy (RE) resources such as biomass, wind, geothermal and solar.

Receiving an average of eight hours of sunlight per day, solar power utilization can be a cheap source of energy in the long run. Solar-power system is indeed a good alternative for power sources especially those in calamity-prone, remote areas and Island provinces of the country.

National agencies such as the Department of Science and Technology particularly the MIMAROPA Region is advocating for the use RE through conduct of forums and technology demonstrations on solar energy. Programs and project of the agency are aligned towards "Green Technologies" (renewable energy utilization) such as allocation of portion of the Regional Grant in Aid budget as leverage for counterpart funds from partner agencies for the procurement of solar panel assembly (Solar Home System). School boards will also be encouraged to have selected schools use solar powered electric fans and serve as demonstration for others and for the students to have a first-hand appreciation of the technology.

The long term plan is to convince Local Government Units (LGU's) for legislation of appropriating at least 2 to 5 per cent of yearly budget for projects related to harnessing available resources for power generation and providing solar-powered facilities for evacuation centers for disaster mitigation. Based on 2012 Internal revenue Allotment (IRA) of 3.3 Billion (U.S.$74 Million) for MIMAROPA Region, 66 Million (U.S.$1.5 Million) up to 166 Million (U.S.$3.7 Million) can be programmed for specific projects like provision of Solar Home System for their constituents- investing today for energy requirement of the future.

Keywords: *Renewable energy, Solar home System, Green technologies, DOST-MIMAROPA, Local government unit, Internal Revenue Allotment (IRA).*

1. INTRODUCTION

The Philippines is considered to be in the typhoon belt receiving an average of 20 typhoons per year. One of the latest is typhoon Yolanda, internationally known as "Haiyan" which heavily devastated the Provinces of Leyte and Tacloban leaving a death toll of 6,115 and about 24.5 billion pesos in damages.

The Philippines has 17 regions and one of the newest which was established in 2003 is the MIMAROPA. It comprises of the island provinces of Mindoro Occidental, Mindoro Oriental, Marinduque, Romblon and Palawan. Cost of electricity and fuels in these areas are higher compared to other regions due to high inter- island transport cost.

Different programs and projects for the utilization of renewable energy resources are being implemented throughout the country most especially for the remote areas and Island provinces like that of the MIMAROPA (Mindoro Oriental and Occidental, Marinduque, Romblon and Palawan) to address issues on poverty reduction and social equity. Rural electrification using renewable energies could be the enabling factor to channel development up to grassroots level of the community.

The Department of Energy reported that demand for solar and wind energy sources is foreseen to grow with the implementation of the program to invigorate the market for solar water heaters and locally fabricated solar panels, dryers and wind pumps. The household sector will remain the largest user of these energy forms.

Realizing this scenario, the Department of Science and Technology thru its regional office for MIMAROPA provinces initiated different programs and projects to mitigate the effects of those problems and prepare its people to be more resilient. It include promotion of biogas technology and use of rice husk gasifier, consultancy services for energy audit and provision of energy efficient production equipment through Small Enterprise Technology Upgrading Programme (SETUP).

DOST-MIMAROPA is advocating for budget allocation from Internal Revenue Allotment (IRA) of each provinces for the adoption of Solar Home System for more efficient and stable supply of electricity especially those in calamity-prone and remote areas.

2. DISCUSSION

Facts and Figures about the Philippines

The Philippines, officially known as the Republic of the Philippines is a sovereign island country in Southeast Asia situated in the western Pacific Ocean. To its north across the Luzon Strait lies Taiwan; west across the South China Sea sits Vietnam; southwest is the island of Borneo across the Sulu Sea, and to the south the Celebes Sea separates it from other islands of Indonesia; while to the east it is bounded by the Philippine Sea and the island-nation of Palau. At 300,000 square kilometres (115,831 sq mi), the Philippines is the 64th-largest country in the world, consisting of an archipelago of 7,107 islands that are categorized broadly under three main geographical divisions: Luzon, Visayas, and Mindanao.

Its location on the Pacific Ring of Fire and close to the equator make the Philippines prone to earthquakes and typhoons, but also endows it with abundant natural resources and some of the world's greatest biodiversity. As reported by DOST PAGASA, the country is visited by an average of 20 typhoons. And with the eminent effect of climate change, the typhoons are getting stronger. One of the latest is typhoon YOLANDA internationally known as HAIYAN, the strongest storm to ever make landfall with wind speed of over 200 km/hr devastating eight (8) provinces in three(3) regions. Among the heavily hit were the province of Leyte and Tacloban city. The typhoon has left an estimated 6,155 dead, 890,895 families or 4,095,280 people displaced and suffering from lack of shelter, food, income and livelihoods. An estimated P 24.5 billion pesos or US$ 533million dollars cost of damage was reported by the National Disaster and Risk Reduction Council (NDRRMC). Electrical and communication systems were shut off for more than a month adding difficulty to rehabilitation and rescue operations.

Population in Philippines increased from 1990 to 2008 by approximately 28 million, a 45 per cent growth in that time frame. With 1.89 growth rate, the 2013 population is estimated to reach 98.7 Million. The population's median age is 22.7 years with 60.9 per cent aged from 15 to 64 years old. Life expectancy at birth is 71.94 years, 75.03 years for females and 68.99 years for males.

The country is divided into 17 regions. These are Regions 1 to 13, National Capital Region (NCR), Cordillera Administrative Region (CAR), Autonomous Region in Muslim Mindanao (ARMM) and MIMAROPA region.

Household Energy Consumption

The 2011 Household Energy Consumption Survey (HECS) revealed that, electricity remains as the most common source of energy by households in the Philippines. About 87 percent of 21 million households used electricity from March to August of that year. About 74 percent of households used electricity while the remaining 26 per cent used kerosene for lighting purposes. Electricity was also popularly used for recreation and space cooling by the household with electricity.

That figure alone suggests that there is a big slice of the pie that can be filled up by renewable energy; solar home system.

With an average cost of P9.00 to P10.00 or US$0.22 per kilowatt hour (kwh) and the increasing cost of electricity, more and more people are now in search of affordable energy solutions. Unlike before that the focus of the problem is on how to have a power supply, the trend now is shifting to cheap, sustainable and reliable source of energy.

Philippines Renewable Energy

The Philippines is the 2nd largest producer of geothermal energy in the world and a wind power producer in Asia with 33MW wind turbines in Bangui, Ilocos Norte. But there is big gap on energy self-sufficiency by about 55 per cent based on the report of Mr. Arnel Antonio of the Department of Energy in 2009.

The country is pushing for energy security with the passing of renewable energy Act of 2008; The enactment of RA 9513, the act promoting the development, utilization and commercialization of renewable energy resources. This law is made to accelerate the exploration and development of renewable energy resources and establish the necessary infrastructure and mechanism to carry out the mandate provided in the law.

One of the laudable projects on Renewable Energy was the BP Solar: Municipal Solar Infrastructure project in partnership with Australian government implemented in 11 provinces in Visayas and in Mindanao. The project uses solar energy as an "enabling technology" to target specific needs and upgrade basic facilities in remote un-electrified communities. A total of 1,145 packaged solar systems were installed in 435 barangays. Community facilities like rural health centers, schools, barangay halls were also provided packaged solar system.

The MIMAROPA Region

The acronym MIMAROPA stands for the provinces of the region, namely, Mindoro Occidental, Mindoro Oriental, Marinduque, Romblon and Palawan. There are two cities in this region, Calapan City in Oriental Mindoro and Puerto Princesa in Palawan. Among the region's provinces, Palawan is the biggest, accounting for almost 54.7 per cent of the regions' total land area while Marinduque is the smallest with an area of only 944.7 square kilometres or barely 3.4 per cent of the total land area of the region.

MIMAROPA's strongest potentials are on agriculture and ecotourism. Forty per cent of the Gross Regional Domestic Product comes from Agriculture and Fishery. It also intensifies food production not just for the region but for the CALABARZON and Metro Manila areas as well.

Among the regions of the country, MIMAROPA remained to have the biggest number of un-energized households as of December 2012 mainly because of the poor access to transportation and the numerous Islets of the region.

The Department of Science and Technology (DOST)

The Department of Science and Technology (DOST) is the premiere science and technology body in the country charged with the twin mandate of providing central direction, leadership and coordination of all scientific and technological activities, and of formulating policies, programs and projects to support national development.

In 1958, Congress passes a law establishing the National Science Development Board (NSDB) upon the recommendation of Dr. Frank Co Tui, who was tasked to survey the state of Philippine S&T during the Garcia administration. It was then revamped in 1982 as the National Science and Technology Authority (NSTA) and accorded broader policy-making and program implementing functions. It was then elevated in 1987 to Cabinet level and becomes the DOST in response to increasing demands for S&T intervention in national development.

Table 7.1: Status of Rural Energization in MIMAROPA

Electric Company	Potential Households	Per cent Energized	Per cent Un-energized
Lubang Electric Cooperative, Inc.	8,900	81	19
Occidental Mindoro Electric Cooperative, Inc.	82,600	77	23
Oriental Mindoro Electric Cooperative, Inc.	169,500	90	10
Marinduque Electric Cooperative, Inc.	49,800	91	9
Tablas Electric Cooperative, Inc.	34,300	85	15
Romblon Electric Cooperative, Inc.	19,000	77	23
Busuanga Electric Cooperative, Inc.	17,800	55	45
Palawan Electric Cooperative, Inc.2	178,000	58	42
Regional Average	**559,900**	**76**	**24**

(http: nea.gov.ph/status-of-energization/category/31-2012-month)

The **DOST-MIMAROPA** is the youngest among the 17 administrative regions of the country. It was created as a separate region through Executive Order 103 issued on May 17, 2003 to accelerate the social and economic development and improve the delivery of public services.

In September 2007, DOST Secretary Estrella F. Alabastro signed a Memorandum separating the management of DOST-MIMAROPA from DOST-IV with a common administrative personnel sharing with DOST-CALABARZON. DOST-MIMAROPA holds office at DOST Complex in Bicutan, Taguig City with its first Regional Director, Dr. Ma. Josefina P. Abilay.

DOST MIMAROPA Banner Programme

Small Enterprise Technology Upgrading Programme (SETUP) was launched in response to the President's call for more focused programs of assistance for Micro, Small, and Medium Enterprises (MSMEs). It is a nationwide strategy to encourage and assist MSMEs to adopt technological innovations to improve their operations and thus boost their productivity and competitiveness. The program enables firm to address their technical problems through technology transfer and technological interventions to improve productivity through better product quality, human resources development, cost minimization and waste management and other operations related activities. The program focused on 5 industry sectors, namely: Food processing, furniture, gift-decors-housewares (GDH), aquatic resources and horticulture and metals and engineering.

The DOST MIMAROPA also provided assistance to State Colleges/Universities, Local Government Units and organized groups through the DOST Grant-in Aid (DOST-GIA). Beneficiaries are provided with financial grants to S&T programs and projects to spur and attain economic growth and development by harnessing the country's scientific and technological capabilities.

In 2013 the DOST MIMAROPA assisted 34 and 18 projects under SETUP and DOST-GIA, respectively. About 50 per cent of that is categorized under food processing sector. Although those projects are mostly micro enterprises, it provided employment to at least 300 individuals per month.

With the uniqueness of the region as composed of 5 island provinces, it is faced with the reality that cost of fuel and electricity is higher than other provinces due to transportation cost. The DOST MIMAROPA realized the need to mitigate that scenario by focusing on projects related to renewable energy.

Under the DOST GIA program, MIMAROPA funded promotion and adoption of Biogas technology using animal manures both from hogs and chicken. Biogas generated isused as fuel for cooking and generators. Another initiative is the promotion of gasifier using rice husk. With the 550,302 hectares of rice field in the region, the potential of the technology is enormous. Carbonized rice hull from gasifier are then used as soil conditioner for organic farming.

DOST MIMAROPA Plans for RE

With the vision to further enhance the capability of the people and be ready for the foreseen energy problems in the coming years, for year 2014, the DOST MIMAROPA will continue to provide funds to strengthen the promotion of biogas and gasifier as new initiative under SETUP. Utilization of most available and sustainable source of energy- the SUN, Technology on Solar energy : Solar Home System (SHS) will be promoted/demonstrated by setting a system in the Marinduque Provincial Science and Technology Center for its lighting, office equipment including internet connections and cooling system. Forums regarding solar power technology will be conducted in the said building for the participants to see and experience the benefits of the technology. The four (4) other PSTC offices and the Regional office will then be provided with Solar Home System in the succeeding years to lower cost on electrical bills and serve as demonstration facilities. With an average of 8 hours of sunlight, solar energy projects can be a cheap source of energy in the long run.

Also this year, about 3 to 5 projects under the SETUP and GIA will be considered to include solar home system as part of the office commitment towards RE and greening of the environment.

In 2015, the regional office will target to have a mobile van equipped with solar panels and common office and home equipment that can be powered by solar energy as demonstration facility. Different municipalities and schools will be the target audience.

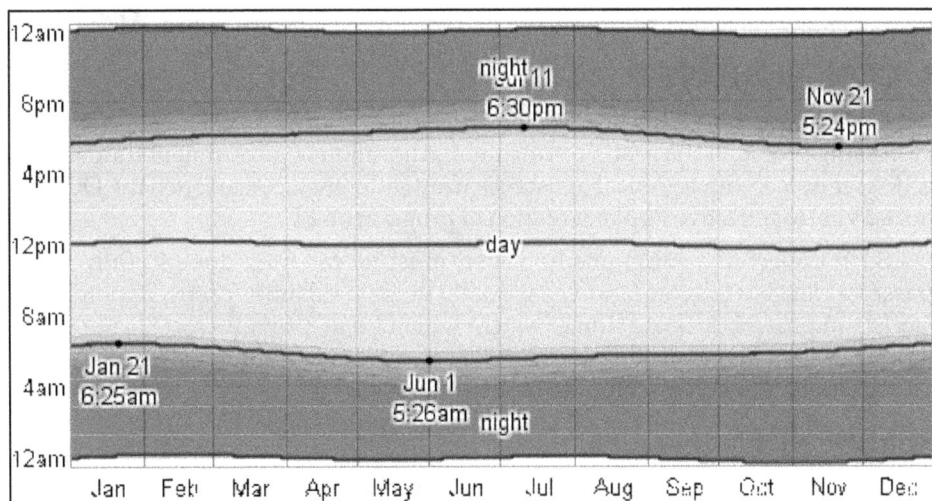

**Figure 7.1: The Solar Day over the Course of the Year 2012. The day, twilights and
night are indicated by the colour bands from yellow to gray
(http: eneken.ieej.or.jp/data/2611.pdf. accessed February 2, 2014).**

The ultimate goal of the DOST MIMAROPA is the institutionalization of RE
budget by the local government units both the municipal and barangay level and the
Department of education to have a yearly allocation from their yearly Internal Revenue
Allotment (IRA) and school board fund.

Table 7.2: 2014 Internal Revenue Allotment for MIMAROPA Provinces

Province	Internal Revenue Allotment (IRA) for 2012	Target 2 per cent of IRA for RE Project		Target 5 per cent of IRA for RE Project	
		Phil Peso	USD at P45/dollar	Phil Peso	USD at P45/dollar
Mindoro Occidental	655,100,000	13,102,000	291,155	32,755,000	727,888
Mindoro Oriental	730,700,000	14,614,000	324,755	36,535,000	811,888
Marinduque	347,800,000	6,956,000	154,577	17,390,000	386,444
Romblon	409,600,000	8,192,000	182,044	20,480,000	455,111
Palawan	1,183,100,000	23,662,000	525,822	59,155,000	1,314,555
TOTAL	**3,326,300,000**	**66,463,000**	**1,478,353**	**166,315,000**	**3,695,886**

http: www.darfu4b.da.gov.ph/background_a.html. accessed February 2, 2014

With full implementation or institutionalization of budget for RE, about
US$739,177.80 funds can initially be used for building and street lighting and for
cooling system projects per year. Since electric fans are the commonly used cooling
system in schools students will have a first-hand appreciation of the solar technology,
it is envisioned that they will have a DEEPER CONCERN and CARE for our
environment.

Let us always remember that every BIG achievement begins with a SMALL step.

3. CONCLUSIONS

Cultivating renewable resources such solar energy is a worthy action in addressing problems in rural electrification. Allocation of government funds (such as IRA) coupled with technical assistance from different agencies such the DOST is indeed vital for effective implementation of program on RE.

Such project and programs however, must be carefully done to fully reach grassroots level of the community especially those in calamity-prone, remote areas and Island provinces. Solar Home System will not only provide electrification and lighting system in the communities but will also provide awareness and education among students and every Filipino on RE initiatives.

REFERENCES

1. http: eneken.ieej.or.jp/data/2611.pdf (accessed February 2, 2014)

2. http: en.wikipedia.org/wiki/Philippines (accessed January 25, 2014)

3. http: region4b.dost.gov.ph/index.php?option=com_content and view=article and id=38 and Itemid=44 (accessed February 2, 2014)

4. http: www.abs-cbnnews.com/nation/regions/11/25/13/yolanda-damage-soars-p245-b (accessed Feb 14, 2014)

5. http: www.census.gov.ph/content/electricity-most-common-source-energy-used-households (accessed February 2, 2014)

6. http: www.census.gov.ph/tags/energy-consumption (accessed February 2, 2014)

7. http: www.darfu4b.da.gov.ph/background_a.html (accessed February 2, 2014)

8. http: www.philstar.com/author/boochanco/demand and supply (accessed February 2, 2014)

9. https://www.doe.gov.ph/renewable-energy-res/biomass-solar-wind-and-ocean

10. https://www.doe.gov.ph/power-and-electrification/expanded-rural-electrification/376-status-rural-electrification/1737-status-of-rural-missionary-electrification

11. http: nea.gov.ph/status-of-energization/category/31-2012-month

Chapter 8

Current Situations and Solutions for Renewable Energy Development in Vietnam

Nguyen Anh Tuan

Ministry of Science and Technology,
Hanoi city, 10000, Vietnam
E-mail: anhtuan@most.gov.vn

ABTRACT

Vietnam has abundant and diversified sources of renewable energy [1], however there is only 1 per cent of potential renewable energy has been used for commercial electricity production. Vietnam is facing several issues regarding to development policy and use of potential renewable energy in order to achieve the target of 3 per cent of commercial electricity produced from renewable energy in 2010, 5 per cent in 2020 and 11 per cent in 2050. While many countries have made renewable energy use compulsory in their national power development strategy, Vietnam remains at the pilot project stage due to lack of capital and technology. This paper analyzes important issues related to national renewable energy policies including legal framework, policies on financial supports, and recommendations for renewable energy development, policies on energy consumption. Current situation and potential of extracting renewable energy is shown: small hydropower, wind power, biomass, solar energy, geothermal energy the basis of following suggestions: develop policies and strategies on renewable energy development in order to promote investment for improving renewable energy sources and consumption; increasing energy commercial price, especially electricity price, is a solution for widely use and exploitation of renewable energy in Vietnam.

Keywords: Renewable energy, Energy policy, Sustainable development in Vietnam.

1. INTRODUCTION

According to Prime Minister's Decision No. 53/2004/QD-TTg on 17th August 2004 about Strategic direction for sustainable development in Vietnam, some types of clean energy which are high potential and can be widely used in household scale such as solar and wind energy, small hydropower and biogas have not been properly studied, applied and generalized; there is a lack of solution for encouraging the use of clean and environmental friendly energy. Therefore, priority activities are: support for research, transfer and application of environmental friendly energy systems including new energy sources and renewable energy sources; encourage the use of less used energy technologies and energy saving programs; preferential financial support for renewable energy development. On 2nd August 2007, Prime Minister made a decision No. 130.2007/QB-TTg about financial policy for investment projects in clean development program (CDM) including projects for electricity generated from renewable energy sources. On 16 June 2009 World Bank signed the agreement on providing USD200 millions for increasing of renewable energy supply for national electricity network in Vietnam [1-4]. Vietnam is going to achieve the target of 3.5 per cent of commercial electricity produces from renewabel energy in 2010, 5 per cent in 2020 and 6 per cent in 2030 [5]. Therefore, the paper shows the current situation and renewable energy exploitation and the current legal framework for renewable energy development; and some suggestions for promotion of renewable energy development in Vietnam. It then proposed solutions for improving renewable energy policy in Vietnam. There's gradual declining in Vietnam country's fossil fuel resources due to the fact that the reserves is limited but the demand is growing and this energy using causes serious environmental pollution. Meanwhile, Vietnam is one of the countries having huge renewable energy reserves in potential. And the development of renewable energy will contribute to reduce not only the fossil fuels spending but also greenhouse gas emission. This article has the following main contents: Overview of Renewable Energy in Vietnam; current situation and renewable energy exploitation and in potential, the current legal framework for renewable energy development; and some suggestions for promotion of renewable energy development in Vietnam.

2. OVERVIEW OF RENEWABLE ENERGY IN VIETNAM

Vietnam is one of the countries having renewable energy source in high potential distributing throughout the country. According to the estimation, biomass in potential from waste products or agricultural waste is about 10 millions tones per year; biogas from garbage, animal manure and agricultural waste is approximately 10 billion m^3 per year; small hydro power(<30 MW) is 4000 MW, and solar power radiation with average sunny radiation is 5 kWh/m^2 per day. Besides, with more than 3,400 km of coast, Vietnam has plentifully wind power source of about 500-1000 kWh/m^2/year in estimation as shown in Table 8.1. These renewable energy sources when being used will meet the rapidly growing demands.

Table 8.1: Current Situation and Potential of Extracting Renewable Energy

Sl.No.	Sources	Reserves in Potential	Exploitation Ability for Electric Production (MW)	Objectives for using
1.	Small hydro power	> 4.000 MW	+ Technology: >4.000 + Economy: 2.200 + Need price support.	Mountainous area: North East; North West, North Central; South Central; Central Highland; Electrical network connection and mini electrial network
2.	Wind power	> 30.000 MW	+ Economy: Need price support	+ The Central, Islands + The coastal and wind areas in other terrains
3.	Solar power	4-5 kWh/m²/day	> 15 MW out of network areas + Need support	+ Solar heat: Residential areas +Solar electricity: Non-residential areas
4.	Biomass +Wood	600-700 MW		Household, small handicraft industries
	+Industrial byproducts		+ rice husks: 197 - 225 + Bagasse: 221 - 276	Rice husk: Mekong River Denta Bagasse: sugar processing areas
5.	Biology +Biogas	> 570 millions m³	58	+ Rural Households + Farms, processing areas
	+Bio-fuels	unidentified	unidentified	+ Transporation + Electricity Production
6.	Geothermal Power	< 400 MW	+ not economical, need support	+ The Central + North East
7.	Tide Power	> 100 MW	unidentified	Littoral areas
8.	Waste	350 MW	222	Urban areas

3. CURRENT SITUATION AND POTENTIAL OF EXTRACTING RENEWABLE ENERGY

Small Hydropower

The extraction of small hydroelectricity in the recent time is estimated to account for about 50 per cent of its potential, of which the rest is located in remote or unfavorable areas. According to recent reports, there are over 1,000 sites that have been identified with potential small hydropower development, scale from 100kW to 30MW with a total installed capacity of 7.000MW. The sites are located mainly in the northern mountainous region, the South Central and Central Highlands.

Wind Power

Considered as a country with potential development of wind power but current data on Vietnam's wind power system has not been fully investigated, due to lacking of exploration and measurement. Data assessment of wind power potential has large fluctuations. According to reports, the wind power potential of Vietnam is most concentrated in the central coastal region, Southern Highlands and Islands.

Figure 8.1: Vietnam Renewable Energy in Potential is Very Plentiful but not yet Estimated in Full–Photographic image

Biomass

Being an agricultural country, Vietnam has huge potential for biomass power. The capicity of sustainable exploitation of biomass resources for power production in Vietnam is estimated to be about 150 million tons per year. The production of bio-fuels has been piloted and commercially produced. In the production of electricity from biomass, some outstanding projects in Vietnam that worth mentioning is thermal power cogeneration from bagasse and rice husks.

Solar Energy

Solar energy for power production comes mostly from solar power cell which are installed in rural, mountainous and island areas. The solar cell system has been in the provinces and cities nationwide. Solar power is used primarily for purposes such as water heating, power generation and other applications such as drying, cooking. With high sunshine hours total to more than 2,500 hours/year, the total average annual radiation around 230-250 kcal/cm² ascending towards the south are good conditions for the development of solar technology.

Geothermal Energy

Despite lacking of a thorough investigation on geothermal, the most recent survey data and assessment indicates potential for geothermal power in Vietnam can be over 300MW. The area with the ability to effectively exploit geothermal energy is the Central.

Thus, there are 5 types of renewable energy has been exploited to produce electricity in our country at present. According to incomplete statistics, the total installed capacity is about 1,215 MW. The renewable energy sources that are being exploited: small hydropower (1000 MW), biomass (152 MW), domestic waste (8 MW), solar (3 MW) and wind (52 MW), which is described as shown in Table 8.2:

Table 8.2: Five Potential of renewable Energy

Line	Sources	Capacity(MW)
1	Small hydropower	1.000
2	Wind power	52
3	Solar power	3
4	Biomass	152
5	Domestic waste	8
	Total	**1.215**

Current situations of renewable energy exploitation is very small compared with the potential, which is accounted for 3.4 per cent. Meanwhile, according to Electricity Planning VII, targets are set to increase the proportion of electricity produced from renewable energy sources accounted for 3.5 per cent in 2010 to 4.5 per cent and 6 per cent in 2020 and 2030, respectively. With the current situation and forecast of the future, Vietnam needs specific solutions to improve the development of renewable energy.

4. THE CURRENT LEGAL FRAMEWORK FOR DEVELOPMENT OF RENEWABLE ENERGY

Policy for renewable energy in Vietnam is operated based on the needs of energy for economic development and environmental protection. As the demand for energy in Vietnam is forecast to rise four times in 2005-2030 and the demand for electricity will increase nine times in 2005-2025, exploiting renewable energy will help reduce reliance on imported energy sources and ensure national energy security for Vietnam. Being aware of the importance of renewable energy, the Government of Vietnam has issued many policies to encourage the development of renewable energy, set the target of using renewable energy and direction to a competitive electricity market with diversified investment and business model. Accordingly, the Prime Minister issued Decision No. 1855/QD-TTg 27/12/2007, approved the National development strategy of Renewable Energy of Vietnam by 2020, with a vision to 2050, and Decision No. 1208/QD-TTg dated 21/7/2011 approved the National plan of development of electricity in 2011-2020, taking the year 2030 in considerartion (master Plan VII), Decision No. 20 177/2007/QD-TTg/11/2007 approved "The project of developing bio-fuels by 2015, with a vision to 2025", in which:

☆ The government has set the target of increasing the share of renewable energy in total primary commercial energy from 3 per cent in 2010 to 5 per cent in 2020 and 11 per cent in 2050 and increase market share of electricity production from renewable energy sources such as wind and biomass from 3.5 per cent of total electricity production in 2010 to 4.5 per cent in 2020 and 6 per cent in 2030.

☆ Regarding bio-fuels, the Government has set a target of achieving an annual output of 100 thousand tons of E5 gasonline and 50 thousand tons of B5 gasoline by 2010, equivalent to 0.4 per cent of total petroleum demand expected for the whole country; 1.8 million tons of ethanol or 5 per cent vegetable oil or petroleum demand in 2025. E5 gasoline is gasoline that contains 5 per cent biodiesel in total volume; B5 is oil containing 5 per cent biodiesel in total volume

To achieve this goal, the government has given many incentives for investors. The renewable energy plants will receive investment incentives, preferential tariffs and tax incentives. Investors can enjoy other privileges such as exemption from import tax and land-using tax exemption for a period of time. Specifically, Joint Circular dated 04/07/2008 58/2008/TTLT-BTC-BTN Environment of Ministry of Finance and the Ministry of Natural Resources and Environment has defined objectives, conditions and method of calculating subsidized rate per unit of production, the amount of annual pension, provision for pension and process of pension request submission for renewable energy projects. However, the current incentives are not enough to form the appropriate conditions for the planning and deployment of renewable energy projects as well as selling the products of renewable energy in Vietnam. The only benefit incentives for small hydropower projects not bring much benefit to other forms of renewable energy.

5. RECOMMENDATIONS AND CONCLUSIONS FOR RENEWABLE ENERGY RESOURCES DEVELOPMENT IN VIETNAM

Based on the above analysis, there are some recommendations as the following:

5.1. Making the Proper Policies of Renewable Energy Resources Development Encouragement and the Reasonable Price Policies

To do this, the policy makers should concentrate on 3 important issues as the following:

Firstly, the policy objectives must be clearly defined, approved by a large number of people, and have flexible mechanisms of promotion while maintaining the stability. This is the basis of implementation of renewable energy price policies as well as many other policies for sustainable renewable energy development.

Secondly, the decision-making process related to renewable energy need to show transparency, accountability and cooperation among whom may it concern. This is important to attract quality contractors for renewable energy resources.

Thirdly, the plan making and implementation of the renewable energy policies must be based on infrastructure and technology corresponding with clear steps from short, medium to long term. These are important factors for not only widespread development of renewable energy technologies but also for the contribution to the development of the energy market with fair competition and reasonable prices for all kinds of energy.

5.2. Building the National Energy Planning

Our country has a huge potential for renewable energy but until now there have not been yet comprehensive studies to know exactly how much it is. Therefore, the planning will help build the full, systematic and reliable data of renewable energy resources using and exploitation potential. From that, we can determine the proper plans of renewable energy resources development in each period and each area.

5.3. Making Specific Plans Implementing Electric Production Programmes from Renewable Energy Resources

It should be provided specific timelines for each period in the roadmap. Initially, it should be focused on the model of electricity power having transmission grid and low voltage of 220. The reasons are this electric power corresponding with the normal devices and easy to regulate the load demands, and that it is more convenient to manage its operation and maintain this electric power. On the other hand, it is necessary to build the service network to provide specialized equipments in the electric system from renewable energy as the controller, inverter and to compile clear documentations for using, and maintenance for the people.

5.4. Increasing Investment in the Basic Researches on Renewable Energy Technologies

An important factor in promoting the development of renewable energy is to increase investment in the basic researches on renewable energy technologies to adopt the most appropriate technologies in renewable energy exploitation and using in each area in Vietnam. Simultainously, by doing this, we can be active in using technologies in renewable energy exploitation and using.

6. CONCLUSIONS

In short, Vietnam country's renewable energy resources is gradually declining because the reserves is limited but the demand is growing. Morever the is serious environmental pollution caused by using this resources of energy. The development of renewable energy is to contribute to reduce the consumption of traditional kinds of energy such as coal, oil, gas and to reduce greenhouse gas emission. Therefore, Vietnam needs to increase investment and support for the development of renewable energy to protect the environment, which ensure an efficient and sustainable development of energy for the country.

REFERENCES

1. Luu Duc Hai, Nguyen Thi Hoang Lien (2009). Renewable energy policies for sustainable development in Vietnam, VNU Journal of Science, Earth Sciences 25 (2009) 133-142.

2. Prime Minister's Decision No. 153/2004/QB- TTg (17th August 2004), Strategic direction for sustainable development in Vietnam.

3. Prime Minister's Decision No. 177/2007/QD- TTg (2(ih November 2007), Project of bio-fuel development to 2015 and the vision to 2025.

4. Prime Minister's Decision No. 130.2007/QD- TTg, Policy on financial issues of investment projects for clean development (CDM).

5. Nguyen, T. (2009), Vietnam yet to develop renewable energy sources, The Saigon Times Daily at http: enplish.thesaigontimes.vn/Home/business/other/4398/ (accessed on 4 December 2009).

Chapter 9

NISIR's Research in Renewable Energy Sources (RES)

Kutemba Kaina Kapanji-Kakoma

National Institute for Scientific and Industrial Research,
Kenneth Kaunda, International Airport Road, Chelstone, Lusaka
E-mail: kkakoma@nisir.org.zm; kainakk@gmail.com

ABSTRACT

The National Institute for Scientific and Industrial Research (NISIR) has been conducting research in renewable energy sources (RES) in the past and is currently running pilot projects in the same. Its research is in line with the Republic of Zambia's country-driven efforts in identifying, prioritizing and addressing technologies that address the key challenges and needs faced in the country, among which include energy needs and the use of renewable and sustainable energy sources for all. Zambia's commitment to efficiently improve energy sources are enshrined in policies and goals such as the revised Sixth National Development Plan (SNDP) of 2013 - 2016, National Energy Policy (NEP) of 2008 and National Technological Needs Assessment (TNA) and Technological Action Plans (TAPs) of 2013, from which excerpts are drawn in the prioritized sectors and technologies. Apart from the sustainable development the renewable energy technologies bring, NISIR has in the recent past engaged in pilot projects that address the co-benefits of renewable energy technologies in the area of climate change mitigation.

The paper will therefore address research currently being undertaken at NISIR and its economic viability in geothermal electricity power generation and bioethanol production using local raw materials and past research in biogas digester technology and its current national status, coal briquetting and improved cook stoves (ICSs) developed by NISIR.

Keywords: Prioritised technologies, Current and past renewable energy sources research.

1. INTRODUCTION

The National Institute for Scientific and Industrial Research is a government institution, set up by the Science and Technology Act No. 26 of 1997 through the Statutory Instrument No. 73 of 1998. This was after the repeal of the National Council for Scientific Research Act Cap 236 of the Laws of Zambia.

Zambia's energy status in terms of national electrification currently stands at about 23 per cent of households, with an urban to rural percentage ratio of 48 to 4 [1 and 3]. Over a decade ago, this figure stood at about 17 per cent with an urban to rural percentage ratio of 44.1 to 2.2 [2].

Currently, Zambia's major renewable energy source is hydro (Table 9.1). The contribution of other renewable energy sources apart from hydropower to the national electricity generation and supply is insignificant [4].

Table 9.1: Installed Generation Capacity in Zambia

Sl.No.	Power Station	Installed Capacity (MW)	Type of Generation	Operator
1.	Kafue Gorge	990	Hydro	ZESCO (Parastatal)
2.	Kariba North Bank	720	Hydro	ZESCO (Parastatal)
3.	Victoria Falls	108	Hydro	ZESCO (Parastatal)
4.	Lunsemfwa and Mulungushi	54	Hydro	Lunsemfwa Hydro Corp. (Private)
5.	Small Hydros - combined	25	Hydro	ZESCO
6.	Isolated Generation	8	Diesel	ZESCO
7.	Gas Turbine (stand by)	80	Diesel	Copperbelt Energy Corp. (Private)
Total Installed Capacity		**1,985 (Estimated Potential of 6000MW)**		

Source: Zambia Energy Sector Profile, Zambia Development Agency, June 2013

Although the potential of renewable energy apart from hydro as a source of electricity generation is high, it remains largely unexploited. Consequently, hydropower (on grid and off grid) remains the major source of electricity in Zambia. There are several reasons why other renewable energy sources have not been intensively exploited. Among them are high investment capital costs, which need guarantees of long-term stable fiscal incentives, such as guaranteed purchase by the Zambia Electricity Supply Corporation (ZESCO), to ensure financial viability. Fiscal incentives and some form of smart subsidies would enable the development of renewable energy sources and make them financially attractive to the private sector's participation [5].

The government of the Republic of Zambia has made many strides in addressing this situation with efforts such as:

1. Developing policy measures and strategies that support sustainable energy sources and state the nation's focus in expanding electricity generation

and transmission capacity, increasing access to electricity and increased utilization of bio-fuels and other renewable energy sources

2. Establishing legal frameworks for energy source management such as Rural Electrification Act, Forestry Act, Energy Regulation Act, Environmental Protection and Pollution Control Act and the Mines and Minerals Act to mention a few

3. Putting in place incentives to attract private sector participation. Some of the incentives on biomass energy, in particular firewood and charcoal, agricultural wastes and forestry waste, which forms the largest part of Zambia's energy mix, is the provision of tax incentives and waivers on biomass energy capital equipment [4]

2. NATIONAL EFFORTS

2.1 National Energy Policy (NEP), 2008

The 1st National Energy Policy (NEP) was formulated in 1994 with a focus on promoting optimal indigenous energy and utilisation for socio-economic development in a safe and healthy environment [4].

It was then revised in 2008 with a more holistic focus on creating conditions that would ensure the availability of adequate supply of energy from various dependable sources at lowest economic, financial, social and environmental costs consistent with national development goals.

Some of the key issues that are highlighted in the reviewed policy include the recognition of the cross cutting nature of energy and takes into account of important issues such as environment and household energy, rural electrification and the role of bio-fuels and Liquefied Petroleum Gas (LPG) in Zambia's future energy mix [4].

2.2 National Development Goals

These national goals focus on inclusive national growth for the sustainable management of the country's wealth as well as the improvements of its citizens in all spheres of life. These goals are all anchored in the country's aim to transition into a middle-income economy by 2030. The formulation of this vision, in 2006, was the first ever written long-term plan. Some of the national development goals are discussed below.

Technology Needs Assessment (TNA) Document, 2013

The government has developed both adaptation and mitigation technologies in most sectors of the economy related to water, sustainable agriculture, energy supply and efficiency and sustainable charcoal production. In addition to this, it has also developed concrete action plans to increase the resilience of the different sectors identified in facing the expected adverse effects of climate change.

The TNA document's initiatives and objectives are to identify and prioritise technologies that can contribute to climate change mitigation and adaptation goals, at the same time meeting the national sustainable development goals and priorities.

It has also been formulated to identify barriers that hinder the acquisition, deployment and diffusion of these prioritised technologies and to develop technology action plans (TAPs) specifying activities and enabling frameworks to overcome the barriers and facilitate the transfer, adoption and diffusion of selected nationally prioritised technologies [5].

The energy sector is one of the identified sectors and it prioritises technologies that address energy supply and energy efficiency like biomass combustion, geothermal, wind energy, biomass wastewater, photovoltaic utility, waste landfill and bio-fuels. Other technologies especially for cooking and heating are listed as improved charcoal stoves, improved biomass institutional stoves, improved firewood stoves, biogas, which can also be used for lighting purposes and solar lanterns [5]. Some of these technologies also have a significant potential of contributing to the reduction of deforestation.

The concepts developed in the TNA are serving as the basis for the development of bankable proposals for financing from various climate change related funding and other bilateral and multilateral arrangements. It is from such prioritised energy technologies that NISIR has sought to engage itself in research on renewable energy for all Zambians.

Revised Sixth National Development Plan (SNDP), 2013–2016

This also considers the critical role different sources of energy play in poverty reduction and national development through its linkages to other sectors of the economy.

In addition, the 2030 long-term national vision aims among other things to have well developed, managed, reliable and sustainable energy services for the improvement of the quality of life for all Zambians in order to make Zambia a middle-income nation by 2030 [6].

The Zambian government has been implementing some of these goals through efforts such as those relating to the Rural Electrification Master Plan (REMP), whose aim is to provide electricity infrastructure and increase electricity access to rural communities. Government is also promoting the use of bio-fuels and has completed its studies and pronounced national blending ratios for bioethanol at 10 per cent (E10) and biodiesel at 5 per cent (B5).

3. NISIR'S RESEARCH IN ALTERNATIVE ENERGY SOURCES

3.1 Current Research: Geothermal Energy Explorations

NISIR and other local stakeholders are collaborating with Kalahari GeoEnergy Ltd, a private company to locate and mine heat from appropriate aquifers in Zambia and if viable develop geothermal energy resources to produce electrical power that could be connected to the national grid or as an off-grid system. Kalahari GeoEnergy Ltd and NISIR have so far conducted exploration drilling.

Geothermal energy is a stable energy source, which is not affected by changing weather patterns and has high efficiency with a thermal capacity potential of 70 – 90 per cent of installed capacity [5]. Some areas of Zambia are part of the East African rift system and have similar hydrothermal manifestations like Tanzania, Malawi and Kenya, which has a more advanced geothermal energy exploitation programme and is currently producing geothermal electricity of around 250MW. Zambia's potential is estimated at around 20 MW.

However, the cost of exploration for geothermal energy may impede widespread use of this resource in Zambia.

3.2 Bioethanol Production

Zambia has a wide variety of crops suitable for bio-energy production due to its good climatic and soil conditions [4]. Therefore, this research has involved the production of bioethanol from starch rich local feedstock such as local wild fruits to produce fuel grade ethanol for blending with petrol up to 10 per cent (E10).

Although the use of local fruits as the sole source of bioethanol may pose a challenge, these local wild fruits can compliment other sources that may be readily available and commercially viable such as sorghum or molasses. Similarly, setting up bioethanol purification plants is initially capital intensive hence would require issues of pricing, cost effectiveness, sustainability and even storage addressed.

3.3 NISIR's Past Research

3.3.1 Biogas Technology for Heating and Lighting

From the early 1980s, NISIR has been studying biogas production, mainly floating dome digesters. Experimental biogas digesters were constructed in various locations, particularly in Southern and Central Provinces of Zambia. The construction of demonstration biogas digesters encompassed a number of rural schools to assist students conduct science laboratory sessions and for lighting purposes.

Many challenges have been experienced by NISIR in trying to advance the adoption of the biogas technology and so this technology still remains in its infancy. The cultural setback associated with producing cooking gas from waste, the lack of understanding of the benefits of the technology, the lack of ownership by the local communities, and the high prohibitive cost have been some of the barriers identified. To date, very few investors have taken on this technology further especially for electricity generation due to the relatively higher electricity generating costs compared to other traditional energy supply systems. However, one notable Non - Governmental Organization (NGO), the Water and Sanitation Association of Zambia (WASAZA) has been using fixed dome digesters and has in some instances constructed them together with the decentralized wastewater treatment systems in schools, hospitals, prisons, farms and abattoirs. In one area, among the feedstock for the "WASAZA" digesters, is the evasive alien species (the Kafue water hyacinth). This pilot project is currently assisting in the clearing of weed - clogged water bodies [7].

Biogas energy for cooking and heating has potential in Zambia. However, the challenge is that generating electricity from biogas has much higher operating costs compared to other traditional energy supply systems and so the cost of the electricity produced would be equally higher.

Exploring avenues such as biomass combustion or biomethanation for on grid and biomass gasifiers for off grid electricity generation is another possibility in the alternative energy mix.

3.3.2 Briquetting and ZIKO Improved Cook Stoves

The Institute with the assistance of the Japanese International Cooperation Agency (JICA) has successfully developed briquetting technology together with the improved cook stoves. The raw material for the briquettes was coal-mining waste products. This technology can be applied to other materials such as sawdust, charcoal powder and waste agricultural biomass.

The institute produced *Ziko* insulated stoves that are more energy efficient compared to the traditional *Mbaula*. They were initially designed to burn the coal briquettes but have now been modified to use ordinary charcoal. Even with the ordinary charcoal, their consumption of the charcoal is still lower than the ordinary ones.

The briquetting and improved cook stoves programme included the training of artisans and currently, these artisans are producing the stoves at a small-scale level.

3.3.3 Participation in Climate Change and other Energy Initiatives

The institute has in the past been involved in the measuring, reporting and verification of pilot projects for their GHG emission reduction potentials. Recently, NISIR was used as a verifier on the Joint Credit Mechanism (JCM) pilot project on green house gases (GHG) emission reduction by the implementation of improved cook stoves in the Central Province of Zambia.

The National Institute for Scientific and Industrial Research currently chairs the Designated National Authority (DNA) technical working group that reviews potential Clean Development Mechanism (CDM) projects. It is also the current Secretary of the Nationally Appropriate Mitigation Actions working group (NAMA WG) under the United Nations Development Programme (UNDP) Low Emission Capacity Building (LECB) project.

Lastly but not least, NISIR also sits on the Renewable Energy Sub-Committee of the Energy Sector Advisory Group (SAG) under the Ministry of Mines, Energy and Water Development.

4. CONCLUSIONS

The National Institute for Scientific and Industrial Research (NISIR) is committed to conducting research that is relevant to the needs of industry and the nation as a whole, in line with the Republic of Zambia's country-driven efforts to improve the lives of every Zambian. Despite it being a publicly funded research institution, it has significantly contributed to the development of different products locally and internationally, and is ready to work with all stakeholders towards the improvement of people's lives, even in making a reality such efforts as sustainable energy for all Zambians.

REFERENCES

1. Banda Chandipo A., Energy Officer, Department of Energy (DoE), Post Newspaper, SE4ALL "Sustainable Energy for All" workshop, 18ᵗʰ November 2013.

2. Census of Population and Housing, 2000, Housing and Household Characteristics, Analytical Report, Republic of Zambia, Central Statistical Office.

3. Living Conditions Monitoring Survey Report, 2006 and 2010, Published March 26, 2012. Republic of Zambia Central Statistical Office.

4. National Energy Policy, Ministry of Energy and Water Development, May 2008.

5. Technology Needs Assessment and Technology Action Plans for Climate Change Mitigation, Zambia, March 2013.

6. Vision 2030, Republic of Zambia, December 2006.

7. Water and Sanitation Association of Zambia website, http: www.wasaza.org.zm/projects.html © 2012, Accessed 13th February, 2014.

Chapter 10

Alternative Energy Utilisation by Institutions of Higher Learning: A Case study of Zimbabwe

Caleb Mharapara

Ministry of Higher and Tertiary Education Science and Technology Development, Zimbabwe
E-mail: calmhara06@gmail.com

ABSTRACT

In the context of energy crisis due to rapid human settlement, high industrial energy demand against the backdrop of declining energy sources and low energy generation capacity, energy supply is going to worsen in Zimbabwe and the SADC Region. The World Solar Summit held on 16th September 1996 in Harare had warned of this impending critical shortage. In this aspect many solar energy applications have been applied at small scale and household level with plausible success. In 2012, a Solar Water Heating System pilot project was installed at one university. Never before was a centralized solar water heating system applied to service a large population in the country. The electricity load shedding affected real time in education and training in Zimbabwe as from 2008. The Ministry of Higher and Tertiary Education was interested in direct benefits on education and training of such an application. A parallel research was carried out during the two year test run of the project to find out the impact and sustainability it would have in institutions of higher learning and its suitability in education and training. The results were found to be multifold, ranging from need for more energy resources for teaching and learning, policy influence on other renewable energy developments and technologies, potential to influence positive perceptions on solar energy technologies by the younger generation, opening new research areas to young scientist and increased usage by industry in the country.

1. INTRODUCTION

Energy is playing a critical role in human development and growth of industry. It is becoming a key factor in any aspect of development and as such a key element in education and training. Educational institutions are playing a key role in the ranking of countries in literacy and development. Institutions of Higher learning have become incubators of great industrial innovations and creation of new knowledge. Sensitive equipment and machinery in the development of medicine, physics, chemistry and many emerging sciences are housed in these centres. The demand for knowledge has risen sharply since the 1970s and huge student communities now reside in Universities and other centres of Higher Learning. Due to these factors demand for electricity to these centres have increased with time. It has become therefore imperative that an uninterrupted supply be put in place to make sure that education and training progresses smoothly without jeopardising quality of the graduates produced. In the face of dwindling energy sources and generation capacities many countries are applying alternative energy initiatives to drive critical sectors of society. Industrial development is directly linked to energy supply and education and training. But due to climate change there is now reduced energy generation capacity especially in Southern Africa where the main energy source is hydro-electricity. Although coal and uranium reserves are still very high their exploitation for energy production is under threat due to international pressure to reduce greenhouse gases. It has therefore become imperative that clean, cheap and locally available alternative sources be researched, developed and used for development especially in the developing world. This will safe guard the continued growth of Education and Training, in itself, a key component of human development. Although Solar is the cleanest and most abundant in Southern Africa its uptake as an alternative energy source has been very slow.

2. METHODOLOGY

For the purposes of this study 8 universities, 4 polytechnic and 3 Teachers Colleges were sampled, and a survey was done in the 15 institutions to find out the status of their energy consumption, usage of alternative energy and measures being put in place against power blackouts. Interviews with heads of the institutions were also carried out. Four site visits were made to assess the state of renewable energy generation. A desk study of the Southern Africa Power Plan (SAPP) and the Zimbabwe Electricity Supply Authority (ZESA) gave the status of the regional and national energy requirements.

Study Objectives

The main objective of the study was to:

☆ Assess the use of alternative energy sources in institutions of Higher Learning in Zimbabwe

Regional Energy Situation

In 1996, Zimbabwe, hosted the World Solar Summit where Heads of states and governments from 18 nations and delegates from more than 100 countries gathered to adopt a Declaration on Solar Energy and Sustainable Development and to launch a 10-year campaign to implement some 300 renewable energy projects around the world. The World Solar Summit culminated a three-year preparatory process initiated by UNESCO to help promote research, education, public awareness, marketing and job creation in all fields of renewable energy. The summit also signalled the start of the World Solar Programme (1996-2005) to implement an initial list of 300 high-priority projects in more than 60 countries. These included solar power schools in rural Georgia, a solar water desalination plant for the Gaza Strip to be implemented by the Palestinian Authority, along with energy self-sufficient homes, farms and villages in Pakistan, Yugoslavia, the Russian Federation and other countries. In Zimbabwe 300 Households were electrified by solar power (Scientific and Industrial Research Development Centre, Zimbabwe 2010). The World Solar Programme 1996-2005 was to be carried out through existing ministries and institutions financed by private and public sources. Major partners include the European Commission, the United Nations Industrial Development Organization, the International Energy Agency, the International Solar Energy Society, the UN Development Programme, the World Health Organization and the World Meteorological Organization.

Up to the year 2000 many of the energy projects planned had not been realised and the region began to experience serious energy shortages. Realizing the danger the energy crisis would have on the region's economic growth, energy ministers from Southern African Development Community (SADC) resolved at a meeting in Harare, Zimbabwe, in mid 2007, as a short-term measure, to undertake generation projects that would add 6,700 megawatts (MW) of power to the regional network administered by the Southern African Power Pool (SAPP) Coordination Centre in Harare. The ministers undertook to raise US$7.88 billion for this project. In addition, the region has a long term project that should come into effect by 2020. The member states wanted to undertake power generation projects, which would add 32,000 MW to the SAPP network at a cost of US$32 billion, according to a statement by the energy ministers, Southern Africa has a current installed capacity of 53,000 MW but the dependable capacity is only about 41,000 MW. The region did project that it would run out of surplus generation capacity by 2007 long time ago but very little was done to address the situation and ensure it had enough electricity beyond that date.

The SADC Protocol on Energy (2010) acknowledges the importance of energy in pursuit of the vision of SADC of economic well-being and poverty eradication in Southern Africa. In order to best achieve these ends, the Protocol on Energy invites Member States to cooperate on energy development, harmonising policies, strategies, and procedures throughout the region. It also advises that these policies ensure the security, reliability, and sustainability of the energy supply, with Member States cooperating on research and development of low-cost energy sources applicable to Southern Africa (SADC Protocol on Energy, 2010).

After this summit only 25 per cent of the projects were really actualised. Why this slow uptake of solar energy in Africa, Southern Africa, in particular Zimbabwe has never been fully explained. Some of the answers may lie in the findings of this paper.

Zimbabwe Energy Requirements

Zimbabwe generates 940 megawatts (MW) of the 2500 MW its population requires. This means that Zimbabwe is dependent on electricity imports from DR Congo, South Africa and Zambia for over half of its total electricity supply. Domestic energy supplies come mainly from the hydro-electric power station and at Kariba, and thermal power stations which have either been shut down due to lack of maintenance.

Although only 40 per cent of the population is connected to the national grid, 62 per cent of Zimbabwe's energy goes to residential use, 9 per cent of energy is spent within agriculture, industry uses 13 per cent and transportation uses 10 per cent of Zimbabwean energy.

The main sources of energy used in Zimbabwe comprise of coal, fuel wood, electricity and petroleum fuels. According to the latest (2000) national energy balance, fuel wood provides the bulk (53 per cent) of the total energy supply, followed by coal (20 per cent), liquid fuels (14 per cent) and electricity (13 per cent). Access to electricity is estimated nationally at nearly 40 per cent, but access to electricity in the rural areas of the country is much lower, at about 19 per cent. This is to be contrasted with the African continent average of around 17 per cent (IEA, 2000). South Africa has a national grid electrification rate of 66 per cent, Zambia 20 per cent and Mauritius 100 per cent (AFREPREN Energy Data, 2004). Rural communities meet 80 – 90 per cent of their energy requirements from traditional fuel mainly fuel wood and 15 – 30 per cent of urban households use wood as the main cooking fuel. Coal, charcoal and LPG are used by very few (<1 per cent) households (Central Statistical Office, Zimbabwe 2002).

The use of fuel wood as an energy source is not sustainable as the rate of use exceeds the natural annual yield. Zimbabwe is currently facing a shortage of electrical energy due to generation shortfalls and has to import more than 35 per cent of its electricity requirements from neighbouring countries (Mozambique, South Africa and Zambia) and the Democratic Republic of Congo. Power imports now constitute a significant foreign currency outflow and have put a strain on the foreign currency situation. Zimbabwe has also to import all its petroleum fuels.

The Zimbabwean government upon realizing the increasing energy demand as settlements increased, two institutions were set up with specific objective to solve this problem. These are the Energy Technology and Scientific and the Industrial Research and Development Centre (SIRDC). The mandate of the two institutions was to:

☆ Spearhead technology development within the energy sector

☆ Adapting, upgrading and transferring of energy technologies for the benefit of industry in Zimbabwe

☆ Promoting efficient use of energy resources

☆ Maximizing energy availability, accessibility and affordability

☆ Spearheading the introduction of environmentally clean and friendly energy resources to the local market

☆ Providing information on energy technology to various stakeholders

Renewable Energy

Research has shown that the country has enormous energy resources which far exceed energy requirements (SIRDC-ETI 2010). Most of these resources are grossly underutilized, particularly natural gas and hydro resources in addition to abundant solar energy throughout the whole year. Solar insolation measurements have shown that Zimbabwe enjoys an average of over 5 hours of sunshine per day. Thus the problem is not lack of energy resources but its development and utilization. The country can become a major net exporter of energy for several decades to come if the energy sources are properly harnessed. In dealing with its energy problems, Zimbabwe faces a unique set of challenging initial conditions, defined mainly by its level and pattern of economic growth, political, social and demographic characteristics, energy resource endowment, location distances between supply sources and consumption areas and technological underdevelopment.

An ethanol plant at Chisumbanje is now producing over 500 million litres per annum, enough for blending with petrol. The blending is still at 10 per cent (E10) but government is aiming to gradually increase to 85 per cent (E85). Although some outlets are already selling E85, the uptake is slow as information and knowledge about that level of blending is not adequate.

The country has over 300 functional bio-digesters (none of these are in higher learning institution) country wide. A biodiesel plant has being constructed but is yet to produce fuel for commercial use due to a number of problems ranging from economic to inadequate stock. Jathropha was targeted as the main stock but plantations are yet to produce sufficient stock to sustainably run the plant.

Population

About two thirds (65 per cent), according to (Central Statistics Office Population Census, 2013), of Zimbabwe's estimated population of 13.5 million live in rural areas but the rate of urbanization is increasing. A rate of natural increase of 1.3 per cent per annum was realised over the period 1992–2002. The rapid urban development is stretching energy supply to the limit. The energy serving mechanisms have been instituted to regulate the consumption and balance the demand and supply side.

Education and Student Population in Higher Education

Education development is a high priority in Zimbabwe and the national literacy rate is now 92 per cent. The country now has now 14 universities, 8 Polytechnics and 14 Teachers' Colleges and several other tertiary institutions with a combined enrolment figure of 110 004 (Table 10.1; Ministry Statistical Report, 2012). Of these 25,400 are accommodated in these institutions and as a result, institutions of Higher Learning, have become huge consumers of energy.

Energy Requirements for Institutions of Higher Learning

Technological advances has brought with it many teaching and learning materials that use electricity and critical among them are internet and computers. The real learning time is now versatile and has no specific time frame for the student. Educational gargets now allow the student to engage in studies any time but at expense of energy. Energy is required to keep every computer, tablet and phone on. Besides energy use by equipment, student tend to use a lot of hot water especially during the cold seasons. The student population is sometimes several times higher during the day in some institutions due to parallel programmes being run as short courses.

Table 10.1: Student Enrolment in Institutions of Higher Learning 2012

Universities	65 825
Teachers' Colleges	18 999
Polytechnics	18 482
Vocational Training Centres	1 277
Others	5 421
Total	**110 004**

Source: Ministry Statistical Report, 2012

The comparative Table 10.2 show the electricity consumption and cost in a sample of selected Universities, Teachers' Colleges and Polytechnics

Table 10.2: Electricity Consumption per Month by Institutions of Higher Learning

Institution	Student Enrolment	Consumption in kWh@ US$ 0.128/unit	Monthly Cost US$
University of Zimbabwe	11 975	625000	80 000
National University of Science and Technology	7 042	163 195	25 170
Midlands State university	16 200	218750	28 000
Bindura University of Science Education	5 362	195312.5	25 000
Chinhoyi University of Technology	5 443	75 969	9 724
Great Zimbabwe University	4 481	609 375	78 000
Harare Institute of Technology	1 458	26 220	3 356
Lupane State University	862	10 000	1280
Belvedere Technical Teachers' College	979	19876	2 544
Seke Teachers' College	1 573	11 654	1 492
Hillside Teachers' College	1319	10 978	1 405
Harare Polytechnic	5 317	18 987	2 430
Kwe Kwe Polytechnic	1 701	16056	2 055
Masvingo Polytechnic	1 907	12437	1 592
Bulawayo Polytechnic	4 057	17684	2 264

It was observed that institutions which have heavy machinery, those with multi-campus and those with high student accommodation rate tend to use more energy.

Alternative Energy Usage by Institutions of Higher Learning

In 2011, the Ministry of Higher and Tertiary Education, in collaboration with Ministry of Energy and Power Development with the assistance of a South Korean company put up a pilot 15, 000 litre Solar Heating System at Harare Institute of Technology, the first of its kind in the country and at an institution of Higher Learning. The Ministry of Higher and Tertiary Education, Science and Technology Development was interested in how this plant would benefit institutions of Higher Learning and how many were using alternative energy to augment their energy requirements (Table 10.3).

Table 10.3: Alternative Energy Use by Institutions of Higher Learning

Institution	Solar	Biogas	Ethanol/ Blend	Bio- diesel	Gene- rator
University of Zimbabwe	X	X	✔	X	✔
Midlands state University	✔	X	✔	X	✔
National University of science Technology	X	X	✔	X	✔
Bindura University of Science Education	X	X	✔	X	✔
Chinhoyi University of Technology	X	X	✔	X	✔
Great Zimbabwe University	✔	X	✔	X	✔
Harare Institute of Technology	✔	X	✔	X	✔
Lupane State University	X	X	✔	X	✔
Belvedere Technical Teachers' College	x	x	✔	x	✔
Seke Teachers' College	✔	x	✔	X	✔
Hillside Teachers' College	X	X	✔	X	✔
Harare Polytechnic	X	X	✔	✔	✔
Kwe Kwe Polytechnic	X	X	✔	X	✔
Masvingo Polytechnic	X	X	✔	X	✔
Bulawayo Polytechnic	X	X	✔	X	✔

There are four institutions that have solar energy sources. They are all solar water heaters for individual housing units. Only one has a central solar heating system supplying hot water to over 1 400 students. No institution is using biogas and only one is using biodiesel. Up to 100 litres of biodiesel are being produced per month and is being used to run its vehicles. Although Zimbabwe has a bio-diesel plant that produces over three thousand litres per month, it is not yet certified for sale to the public. All the institutions are using ethanol blend for all their petrol cars (E10).

It was observed that every institution was offering boarding facilities to a number of students and was throwing away over 150 kg each of waste food daily. This has a potential to produce enough biogas for cooking and heating.

All the institutions have diesel generators of various sizes which are used for three to five hours per week. However the generators are running on diesel which is an extra cost.

Table 10.4: Cost of Generator Use in institutions

Institution	Diesel in litres	Amount @ US$ 1.35
University of Zimbabwe	350	473
Midlands State University	7400	9 990
National University of Science Technology	6 714	9 063
Great Zimbabwe University	44 400	59 940
Bindura University of Science Education	2 370	3 200
Chinhoyi University of Technology	6 100	8 235
Harare Institute of Technology	1 000	1 350
Lupane State University	4 148	5 599
Belvedere Technical Teachers College	4 140	5 600
Seke Teachers College	2 678	3 615
Hillside Teachers College	3 400	4 590
Harare Polytechnic	1 780	2 403
Kwe Kwe Polytechnic	5 904	7 970
Masvingo Polytechnic	3 800	5 130
Bulawayo Polytechnic	4 800	6 480

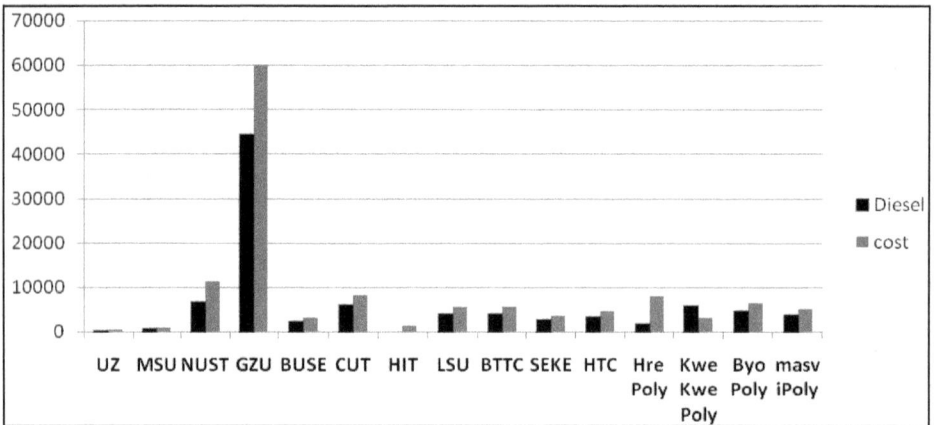

Figure 10.1: Cost of Running Generators in Institutions of Higher Learning

3. FINDINGS

1. Only one institution is using bio-diesel to run its two generators
2. Only one Institution has a Solar Water Heating Systems.
3. Most institutions are not well informed about solar and Bio-gas energy utilisation.
4. Solar panels in Zimbabwe are expensive to install as they are sold as separate units instead of kits.

5. All institutions are throwing away considerable amount of biodegradable waste.

4. RECOMMENDATIONS

There is need to vigorously carry out a campaign in Institutions of Higher Learning in Zimbabawens to start utilising solar and green fuels as alternative energy.

1. Replicate the installed solar heating system to all institutions.

2. Lobby government to waive or reduce tariffs on import of equipment for alternative energy for education.

3. Construct biogas digesters in all institutions to reduce energy demand from the national grid.

4. There is potential for biogas production in all the institutions as they are throwing away considerable amount of biodegradable waste.

5. CONCLUSIONS

The Southern African region has an energy crisis and inadequate energy generation capacity which is not likely to be solved in the near future. The Regional Energy protocols in place may require ten to fifteen years if they are implemented at all. The diesel generators in common use are increasing the energy expenditure for the institutions. There is huge potential for alternative energy generation and utilisation in Zimbabwe. Zimbabwe will need to carry out vigorous campaign awareness on the utilisation of alternative energies in its Institutions of Higher Learning. The graduates from the institutions will then be able to influence on opinion on the use and effectiveness of alternative energies.

REFERENCES

1. Manpower Planning and Development Act (1996)

2. Ministry of Energy Development, Zimbabwe energy Development Policy (2008)

3. Implementation of Renewable Energy Technologies - Opportunities and Barriers, Zimbabwe Country Study". United Nations Environment Programme Collaborating Centre, Risøe, Denmark. 2001

4. SADC Protocol on Energy (2010)

5. SIRDIC-ETI Report (2010)

6. Southern Africa Power Plan (SAPP)

7. Zimbabwe Central Statistics Office Population Census (2013)Nations Environment Programme Collaborating Centre, Risøe, Denmark. 2001

8. www.sadc.int/documents./show/Protocol per cent 20on per cent 20Energy per cent 20(1996)

9. www.sadc.int/./Regional_Infrastructure_Development_Master_Plan_En.

10. www.tradingeconomics.com/zimbabwe/electric-power-consumption-kwh.

11. www.marketresearch.com/Business./Zimbabwe-Power-Q1-72807

12. China-backed firm plans 600 MW Zimbabwe electricity plant, International: Reuters, 2013

13. https://www.cia.gov/library/publications/the-world-factbook/geos/zi.html CIA Factbook: Zimbabwe 2010

Chapter 11

Technology Transfer in the Area of Alternative Energy–Why Zimbabwe Remained Bogged in Load Shedding: A Case Study of Photovoltaic GEF Project Zimbabwe

Ngonidzashe Givemore Dupwa

Science and Technology Officer, Ministry of Higher and Tertiary Education,
Science and Technology Development, Harare, Zimbabwe
E-mail: ngdupwa@gmail.com

ABSTRACT

Zimbabwe was one of the first countries to sign and ratify the UN Framework Convention on Climate Change (UNFCCC) in quick succession, and by doing so confirmed its commitment to the global principle of common and differentiated responsibility in climate change mitigation. Secondly, as one of the founders of the Global Environment Fund (GEF), Zimbabwe was rightly placed to qualify as one of the recipient countries of initial GEF funding. One of the objectives of the project was to enhance and upgrade the Zimbabwean solar manufacturing and delivery infrastructure. This paper will analyse the major players involved in the technology transfer lens. The paper will give the role of stakeholders in the implementation of the technology transfer projects in climate change mitigation using the GEF solar project in Zimbabwe. The intellectual property dimension in renewable energy technologies will also be discussed and the pitfalls that arise from overlooking such a pertinent issue. Successful technology transfer needs to go through the perilous phases of the assimilation and absorption of technological knowledge: adaptation to local conditions, absorption of subsequent improvements and generalisation of the transferred knowledge.

The paper will explore to what extend has Zimbabwe domesticated some of these provisions of the International agreements to increase beneficiation in the energy sector.

1. INTRODUCTION

The TRIPS Agreement is a minimum standards agreement, which allows Members to provide more extensive protection of intellectual property if they so wish. The TRIPS Agreement eventually came into effect on 1ˢᵗ January 1995 of which Zimbabwe ratified the treaty by the 5ᵗʰ March the same year. Developed country Members shall provide incentives to enterprises and institutions in their territories for the purpose of promoting and encouraging technology transfer to least-developed country Members in order to enable them to create a sound and viable technological base. Incentivising foreign firms to enter such transactions is a clear opportunity for developed country governments to properly fulfil their obligations contained in Article 66.2 of the World Trade Organisation (WTO) Agreement on Trade-Related Aspects of Intellectual Property Rights (TRIPS). It is form this background that the emphasis on technology transfers is being discussed with Zimbabwe being a developing country to benefit from the arrangement.

The UN Framework Convention on Climate Change (UNFCCC) is an intergovernmental treaty developed to address the problem of climate change. The Convention, which sets out an agreed framework for dealing with the issue, was negotiated from February 1991 to May 1992 and opened for signature at the June 1992 UN Conference on Environment and Development (UNCED) — also known as the Rio Earth Summit. The UNFCCC entered into force on 21 March 1994, ninety days after the 50th country's ratification had been received. By December 2007, it had been ratified by 192 countries (IISD report, 2013).

Zimbabwe was one of the first countries to sign and ratify the UNFCCC convention before it even entered into force on 21 March 1994 demonstrating its commitment to the principles thereof.

Article 4.5 of the United Nations Framework Convention on Climate Change (UNFCCC) states that developed countries "shall take all practicable steps to promote, facilitate, and finance, as appropriate, the transfer of, or access to, environmentally sound technologies and know-how to other Parties, particularly developing country Parties, to enable them to implement the provisions of the Convention." In this context, technology transfer is designed to assist developing countries with responding to climate change through the diffusion and use of appropriate climate change mitigation and adaptation technologies.

By 1996 the Zimbabwean government had made some progress as far as climate change mitigation strategies and responses were concerned. In its first national communication report four major options had been selected for climate change mitigation. In this paper only the energy related options are going to be considered as follows:

1. Introduction of More Efficient Coal-Fired Industrial Boilers
2. Photovoltaic Technology

Introduction of More Efficient Coal-Fired Industrial Boilers, the main aim of the government was to improve the efficiency to 80 per cent through the introduction of soot blowers to clear steam pipe surfaces and thus increase heat transfer. In 1998, the Government of Zimbabwe received assistance for the Global Environment Facility (GEF) to develop a project to address industrial energy efficiency improvement. The project did not take off because the Zimbabwean counterpart failed to raise enough money to co-finance the project. The modalities of the project involved technology licences between boiler manufacturers largely based in industrialised nation and auxiliary equipment manufactures in the developing country, in this case Zimbabwe (Horta, 2002). The GEF will the pay part of technology licence for the firm in the developing country and the other part will then come from the developing country. In Zimbabwe considering that many companies were using coal fired boiler at the time the companies did not subscribe to the offer which led to the failure of the programme as there were no takers. This may be due to the fact that most companies did not have information on the benefits of energy efficiency to company objectives. In the lens of Intellectual property it also showed that there was a general lack of awareness to the authorities to the benefits that could have been accrued by the country had the companies utilised the opportunity of technology transfer and gained capacities to make efficient industrial boilers. The knowledge acquired would have gone a long way in advancement of technology to the Zimbabwean companies. Another factor that could have led to the lack of appreciation of energy efficiency in Zimbabwean companies was due to energy subsidies that they received from government when using the coal so the companies tended to be reluctant to include energy efficiency on their company objectives.

Photovoltaic Technology

The Government did embark on the project to introduce photovoltaic technology especially in rural areas with a passion. This paper will delve much on this project and discuss the main issues to be connected for through all the multilateral frameworks available and how they were utilized and possible comments suggested for improvement. The main aim of the project was to reduce the use of firewood in the rural areas as stated in the first National communication. Applications of solar technology in Zimbabwe have been through donor supported programmes which have focused on production of electricity for pumping water, refrigeration, lighting in households and institutions, powering of radios and TVs as well as solar thermal (heat production) for cooking, water heating and crop drying. It was a typical technology transfer project in there frameworks as will be shown.

Solar PV Projects in Zimbabwe

Table 11.1: Solar PV Projects in Zimbabwe

Donor	Year	Area of Support	Total Amount
UNDP GEF	1993-1998	Solar PV programme	USD7 million
JICA	1997-1999	Installation of cluster PV systems	USD10 million
German Technical Cooperation	1992-1994	Solar water pumping	USD 6.5 million
Chinese government	1999	110 solar PV systems and water pump	
Italian government		Lighting of rural schools	USD 92000

One of the objectives of the solar PV was to train and human capacity building in Zimbabwe. The PMU was expected to launch a series of training programmes designed to develop a critical mass of locally trained personnel in component manufacturing, and system installation and maintenance. The PMU also had plans to train end-users in basic maintenance.

Lessons Learned

The objective of training and human capacity building in solar PV project was not met as evidenced that at the end of such projects. The participants did not only remain doing component manufacturing. This was due to the fact that the training was program based and not from the institution dedicated to offering training in such technical areas that is Polytechnics and Universities. Flow of know-how between entities should to be managed by specialists in curriculum, so that the level of assimilation can be ascertained. This would have resulted in a situation where the recipients of such knowledge would have continued with the project well after it is concluded as they may have acquired the art. This was not the main objective as it seemed like the companies that participated just wanted to sell their products as the programme were well funded.

Technology Transfer Definitions

The IPCC report on methodological and technological issues in technology transfer (IPCC 2000) contains a broad definition of technology transfer which they define in terms of a set of processes "covering the flows of know-how, experience and equipment, for mitigating and adapting to climate change amongst different stakeholders such as governments, private sector entities, financial institutions, NGOs and research/education institutions."

According to the IPCC (2000), technology transfer involves the process of "learning to understand, utilise and replicate the technology including the ability to decide which technology to transfer and adapt it to local conditions and integrate it with indigenous technologies."

Bozeman, (2000) in his overview of technology transfer, defines the concept as "the movement of know-how, technical knowledge, or technology from one organizational setting to another".

Technology transfer basically consists of a"transfer" of technological knowledge, including a "transfer" of the capacity to assimilate, implement and develop a technology.

Thus a successful TT goes through the perilous phases of the assimilation and absorption of technological knowledge: adaptation to local conditions, absorption of subsequent improvements and generalisation of the transferred knowledge (Foray, 2009).

Technology transfer at its best can act as a vehicle for change from unsustainable practices and improve quality of life but it is the problem of inducing sustainable change that means that any transfer must be carefully tailored with country stakeholders to meet their needs (Flamos and Begg, 2010). It is from this background of technology transfer that the solar PV project is going to be analysed.

Support Structures for Technology Transfer

Alavi, and Leidner, (2001) pointed out some challenges that need to be addressed if technology creation is not to impede the transfer process includes lack of the following:

1. Information relevant to strategic planning and market development
2. Science, engineering and technical knowledge, especially in the private sector
3. Institutions with the mandate and resources to equip people with the requisite knowledge and skills
4. Research, development and testing facilities
5. Technology development and adaptation centres
6. Joint industry-government planning and collaboration.

2. POLICY DISCUSSIONS

These sentiments can be echoed among the players involved for the local installation companies that participated in the project. As (Mulugeta *et al.*, 2000) stated that of the 60 companies that were registered for the project in 1997 only 30 of the companies renewed the registration to the Solar Energy Installers Association of Zimbabwe in 1998 when the project was winding up. The implication of such a high level of dropping out by these companies may be that some of the companies never engaged in the solar business in the first place or no longer engaged in the in the PV installation due to industry shake outs. Most of the companies that did not have business visions and remained content with the hand-outs from the project and did not innovate and prepare themselves for post GEF ordeal were too weak to continue operating in PV business.

From a technology transfer perspective Seaton and Hayes, 1993 discovered that the failure by some of these companies may be due to the fact that the transfer mechanisms in the GEF solar project tended to emphasize on the marketing and selling of the technology as products to companies that had explicit needs and request rather than provide a business service that aided the process of diagnosis, searching for matching the available technology to implicit needs. Another important aspect that may have been overlooked in the solar projects in Zimbabwe was the underestimation of the importance of the interactive processes between the donor (vendor, intermediary, R&D organisation etc.) and the recipient, necessary for successful transfer. The project failed to recognise that technology transfer seldom involves just a simple one off transaction but is a process or a dialogue between a diversity of actor in the two parties and involves a continuing relationship to the point where real benefit accrues to the recipient companies.

As in Figure 11.1 there are a lot of missing actors in the technology transfer framework is to be complete. Since some of the companies had been formed based on the inventions in this case the solar PV project it was required to have incubators. The incubator could have increased the survival rate of the innovatory technology as it had just emerged from the research world into Zimbabwe. The presence of company incubators would have enhanced the opportunities for networking, providing tenants (companies) with the appropriate technical and other support infrastructures and services (Salvador, 2010). Entrepreneurial dynamics constitute an important mechanism to regional development, whether deriving from academic spin-offs, rendered support by science and technology parks and incubators, as defended by Salvador (2010); or resulting from the founding and expansion of family based companies, as proposed by NordqvistMelin (2010) and, in either case, resulting in locally produced employment and wealth. It was imperative that to foster the entrepreneurial skills amongst the young companies there was need to include the incubation framework in the solar PV projects in Zimbabwe. It would have gone a long way in training the business persons such that when the project had ended institutional memory on the projects would have need successfully transferred the solar PV technology into Zimbabwe. The incubator would also have helped with the following for the actors involved with capacity building, technology and skill, marketing access, business information and to some extent negotiation with the government. There was need to include an incubation function within the PMU of the solar PV projects in Zimbabwe (Tatikonda and Stock, 2003).

Apart from the incubator a technology transfer center was necessary. In this case the Ministry of energy, Agribank and other actor took over such role. A technology transfer center would have had a representation of the national patent office which was not part of the solar PV project in Zimbabwe. The Zimbabwe Intellectual Property Office (ZIPO) would have advised the actors on the patents involved in the solar technology being transferred and making break ground checks on the newness of the technology. The technology transfer center could also provide a complete and detailed technical description of a desired technology to all the actor involved even conducting a novelty search before a research project started to avoid investment in problems already solved.

Creation of useful and usable technologies is a major factor in ensuring there is the opportunity to make informed and confident choices in technology investment projects.

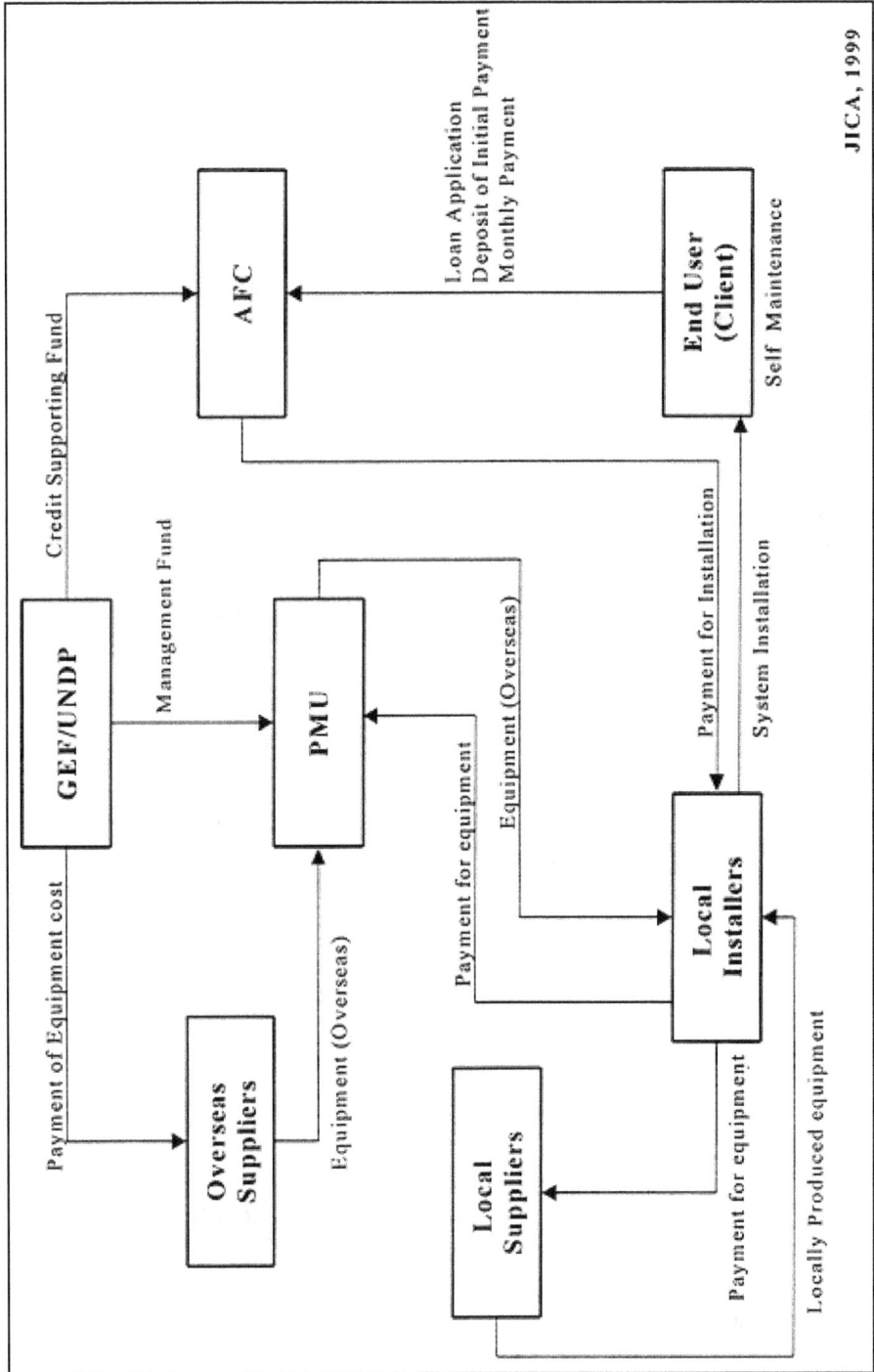

Figure 11.1: The GEF/UNDP Solar PV Programme

JICA, 1999

3. CONCLUSIONS

A national collaborative technology transfer team should have been formed to represent a broad range of interests, including sectoral ministries – such as energy, transportation, agriculture, forestry, and water resources – environmental agencies, climate change delegates, private companies and trade associations, technical institutions, and non-government organizations. The ministries that implement energy and other sector specific programs and policies were particularly important because they often had the controlling or regulatory authority necessary to implement technology transfer actions. Climate officials should have been involved as well, even if they did not have implementation responsibilities, so that they could integrate the work of the team with key climate change issues and identify opportunities to build on other climate change work. Government could have played a key role in facilitating these partnerships. In particular, effective inter-ministry cooperation would have maximized synergies between various programs and avoid conflicts with other government programs and objectives.

REFERENCES

1. Alavi, M., and Leidner, D. E. (2001). Review: Knowledge management and knowledge management systems: Conceptual foundations and research issues. *MIS quarterly*, 107-136.

2. Dechezleprêtre, A., Glachant, M., and Ménière, Y. (2013). What drives the international transfer of climate change mitigation technologies? Empirical evidence from patent data. *Environmental and resource economics*, 54(2), 161-178.

3. Flamos, A., and Begg, K. (2010). Technology transfer insights for new climate regime. *Environment, development and sustainability*, 12(1), 19-33.

4. Foray, D. (2009). *Technology transfer in the TRIPS age: the need for new types of partnerships between the least developed and most advanced economies*. ICTSD International Centre for Trade and Sustainable Development.

5. Horta, K. (2002). The Global Environment Facility the First Ten Years œ Growing Pains or Inherent Flaws?

6. Hübler, M., and Finus, M. (2013). Is the risk of North–South technology transfer failure an obstacle to a cooperative climate change agreement?. *International Environmental Agreements: Politics, Law and Economics*, 1-19.

7. Ikeme, J. (2003). Climate Change Daaptational Deficiencies in Developing Countries: the Case of Sub-Saharan Africa. *Mitigation and Adaptation Strategies for Global Change*, 8(1), 29-52.

8. Lanjouw, Jean Olson and Mody, Ashoka. (1996). Innovation and the international diffusion of environmentally responsive technology, *Journal of Research Policy*. http: www.sciencedirect.com/science/article/pii/0048733395008535

9. Ma, Z. F. (2012). The effectiveness of Kyoto Protocol and the legal institution for international technology transfer. *The Journal of Technology Transfer*, 37(1), 75-97.

10. Tatikonda, M. V., and Stock, G. N. (2003). Product technology transfer in the upstream supply chain. *Journal of Product Innovation Management*, 20(6), 444-467.

SECTION III
Energy Efficient Technologies and Smart Grid

Chapter 12

Monitoring and Novel Applications of 220kV/500kV Egyptian Grid Parameters Using Family of PMU Based WAMS

M.M. Eissa[1]*, Mahmoud M. Elmesalawy[2],
Ahmed A. Shetaya[1] and Ahmed H. Soliman[2]

[1]*Department of Electrical Engineering,*
Faculty of Engineering-Helwan University, Cairo, Egypt
[2]*Department of Electronics, Communications, and Computer Engineering,*
Faculty of Engineering - Helwan University, Cairo, Egypt
*E-mail: mmmeissa@yahoo.com

ABSTRACT

Egypt is moving towards smart grid infrastructures to enable efficient bidirectional power supply with reduced carbon footprint. The target smart grid will feature distributed energy generation with renewable energy systems. With the improvement of wind power technology in Egypt, increase of wind power capacity, the impact of wind power on the grid has become an important research topic as well. Smart Grids are facing new challenges, such as increasing penetration of renewable energy sources, in particular wind generation, growing demands, limited resources, and competitive electricity markets. The technology of synchronized phasor measurements is well established. It provides an ideal measurement system with which to protect, monitor and control a power system, in particular during conditions of stress. System wide disturbances in power systems are a challenging problem for the utility industry because of the large scale and the complexity of the power system. The Egyptian Wide Area Monitoring System (EWAMS) is wide area system is designed in EGYPT. A family of the PMU are deployed on a real 220kV grid. The Synchrophasor system with wide deployment of phasor measurement units and using high-speed communications to deliver and collect synchronized high-speed grid operating data. The system is cable for coping many of real time applications.

Keywords: *Wide-area measurements, Frequency monitoring network, System architecture, 3G, GPS, Monitoring and applications.*

1. INTRODUCTION

Recent developments in smart grid technology have spawned interest in the use of phasor measurement units to help create a reliable power system transmission and distribution infrastructure. Wide-area monitoring systems (WAMSs) utilizing synchrophasor measurements can help with understanding, forecasting, or even controlling the status of power grid stability in real-time. A power system Frequency Monitoring Network was first proposed in 2001 and was established 2004. As a pioneering WAMS, it will serve the entire Egyptian power grid through advanced situational awareness techniques, such as real-time event alerts, accurate event location estimation, animated event visualization, and post event analysis. This project presents implementations of FNET's and the necessary applications as part of the Egyptian Grid with renewable resources. This will add significant capacities to the Egyptian system for observing power system problems. EWAMS architecture is installed. Many of on-line applications will be given such as "Event detection and location", "Visualization", "Detection Changes", and "Island Protection". The following application can be verified through the project "Modeling and Simulation", "Intelligent Algorithms", "Intelligent Tools", "Automatic Synthesis", and Renewable Penetration on the Grid". These complex challenges are driving the evolution of smart grid technologies. The greatest future challenge is the integration with the renewable energy resources and controlling of it to reflect the best impact on the massive energy production [1].

The smart power transmission networks are conceptually built on the existing electric transmission infrastructure. However, the emergence of new technologies (e.g new materials, electronics, sensing, communication, computing, and signal processing) can help improve the power utilization, power quality, and system security and reliability, thus drive the development of a new framework architecture for transmission networks [2]. The initial field installations of the PMUs were in the service territories of Bonneville Power Administration, American Electric Power, and the New York Power Authority. At present, there are only 105 PMUs installed in the Eastern Interconnection (EI) and 56 in the Western Interconnection (WECC). It can provide information about generation electro-mechanical transients, generation-demand dynamics, and system operations, such as load-shedding, break reclosing, and capacitor bank switching. This characteristic allows frequency monitoring to be as informative at the distribution level as it is at the transmission level. Phasor calculation in the FDRs is derived from real-time voltage signal sampling and its Discrete Fourier Transform-based complex representation. Voltage phase angles and magnitudes are also measured by FDRs, which can provide useful information for power system event recognition and status estimation [3].

Smart Grid in EGYPT should be started. The improvement of wind power technology, increase of wind power capacity, the impact of wind power on the grid has become an important research topic as well. Large power systems are facing new challenges, such as increasing penetration of renewable energy sources, in particular wind generation, growing demands, limited resources, and competitive electricity markets. Under these conditions, the large power systems have had to confront some major operating problems in voltage regulation, power flow control, and stability [4]-[7].

This paper describes wide-area synchronized measurement system architecture as a pioneering WAMS for the first time in Egypt to cape with the world in solving the wide area power system from the operation to coordination and then control action. The implemented Egyptian WAMS (EWAMS) consists of 11 FDR units geographically dispersed throughout the boundary of the Egyptian power grid, and a data management center (HUHS) located at Helwan University. EWAMS is implemented to serve utilities, academics, and policy makers with reliable and valuable synchrophasor data from the Egyptian power grid. Based on the valuable information provided by the installed FDRs, a variety of applications was developed in order to assess and improve the performance of the Egyptian power grid during steady state and transient conditions.

2. WIDE AREA MONITORING SYSTEM

Wide-area measurement systems (WAMS) can be defined as a system that takes measurements in the power grid at a high granularity, over a wide area and across traditional control boundaries, and then uses those measurements to improve grid stability and events through wide-area situational awareness and advanced analysis [8]. Certain power system measurements cannot be meaningfully combined unless they are captured at the same time. Many advanced applications can use advantage of the measurement capability provided by WAMS as wide-area monitoring, real-time operations, improved accuracy of models and forensic analysis, in which wide-area measurements such as frequency measurements can be used to provide authentication of different recordings.

A typical wide-area measurement system or WAMS system is built upon a reliable communication system connecting power stations, network control centers and sub stations. Figure 12.1 shows an example for wide area monitoring system. The GPS satellite system is used for synchronization and timing accuracy, and a number of Phasor Measurement Units or PMUs stream the synchronized real-time data through the communication link to data concentrator. In a power system, operators rely on system models to know how power flows will change either as a result of manual or relay-initiated changes to topology, and they can make economical changes at low customer cost. However, if the system model is not accurate enough, such actions may violate the system stability criteria, such as loading a transmission line over the stability limit. Therefore using WAMs to validate a system model and improve model accuracy are very crucial for the economic optimization of a system, and to insure operators are not violating system stability [9].

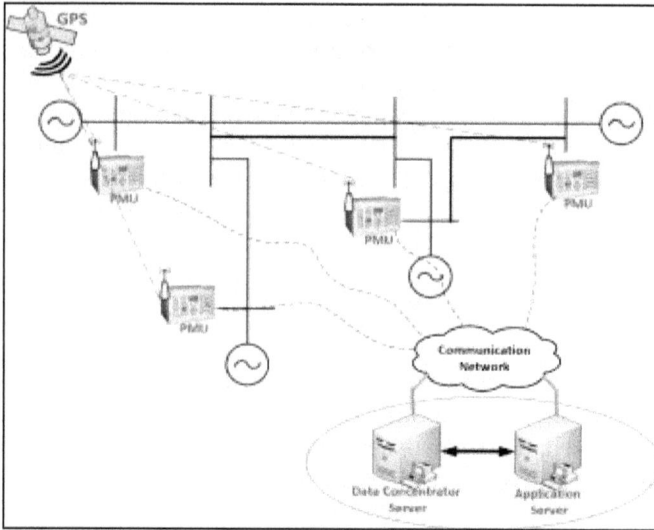

Figure 12.1: Smart Grid with Data Synchronization

WAMs can also be used for power system post-event analysis. Since all the data of WAMs are synchronized and time tagged, using WAMs as the data resource for the timeline is an obvious choice. With its help, the system dynamics during an event can be reconstructed and the sequence of events during cascade events can be revealed. Comparing the measured data at different time points with system model-based simulation, one can disclose what happens in the process of the event and what the impact is [10].

The FDR is a low-cost, quickly deployable frequency measurement device. Its price is much less than that of a commercial Phasor Measurement Unit (PMU), and requires no installation design or fee. Figure 12.2 displays the second generation of the device. The FDR measures the voltage signal from ordinary 100 V, 50 Hz phase-phase system and sends the synchronized frequency and voltage phasor data back to the centralized server via the Internet. It has virtually no algorithm error at 50 Hz and its calculated frequency accuracy is better than ±0.0005 Hz [10]-[11].

The FDR data can be used to enhance grid reliability for both real-time operations and offline planning applications, as listed below; the benefits of achieving such system will assist in:

3. EWAMS ARCHITECTURE

Egyptian Wide Area Monitoring System (EWAMS) is a 3G based wide area sensor network consisting of a special type of high precision family of the phasor measurement units (PMUs) and a central data management and processing system. EWAMS is a smart grid system designed to collect real-time synchronized frequency, voltage, and phase angle measurements at the transmission level 220kV/500kV of the power grid.

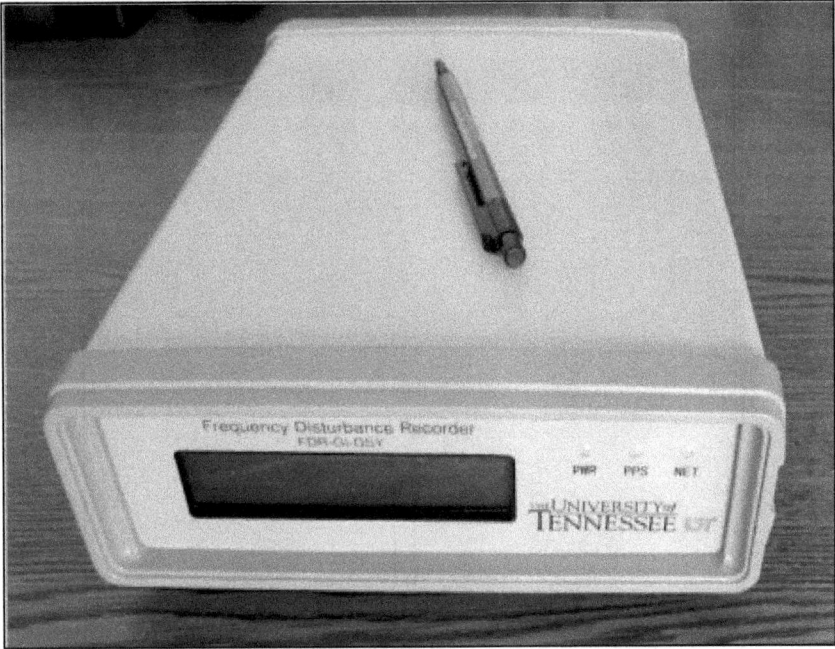

Figure 12.2: Frequency Disturbance Recorder [10]

The structure of the EWAMS can be classified into five levels. The first level represents the placements of power stations on the power grid in which the sensors are installed. The second level contains the GPS enabled sensors that provide frequency, voltage magnitude, and voltage angle measurements. The third level is the communication infrastructure that provides the integrated wide area communication media for data measurements transmission. The fourth level is the remote data management and processing center that provides data gathering, storage, web service, post-disturbance analysis and other information management functions. The last level is the secure remote access connection for different EETC sectors and other remote clients. Figure 12.3 shows the EWAMS architecture deployed for the Egyptian power grid. In order to discuss the unique characteristics of the EWAMS, the main building blocks of the EWAMS are discussed in more details in the following subsections. Figure 12.4 shows the procedure for the remote access designed for the EWAMS system helping the Egyptian Electricity Company for monitoring fast action fo the parameters.

EWAMS Sensors (FDRs)

The Frequency Disturbance Recorder (FDR) is considered as the key component of the EWAMS. It works as a sensor which performs real time synchronized measurements for frequency, voltage magnitude, and voltage angle and transmits these measurements to a remote data center, hosted in Helwan University for management, processing, and long term storage. Each FDR is equipped with a GPS receiver, which is used to provide the accurate time signal needed for synchrophasor

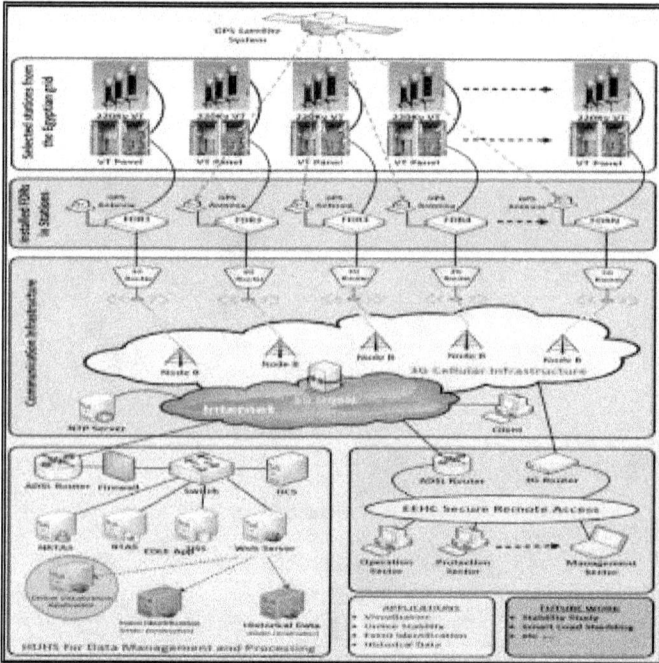

Figure 12.3: EWAMS Architecture

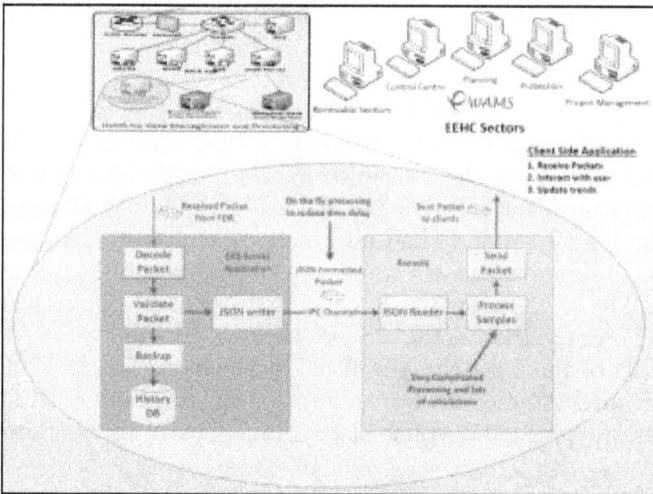

Figure 12.4: Remote Access for EWAMS

calculation. Also each FDR is embedded with 10/100 Mbps Ethernet interface for communication capability.

In EWAMS, the FDR is especially designed and implemented to have two power inputs. The first one is the single phase voltage (57.73VAC) which is taken from the 220/500 kV voltage transformer output in each power station and is used for estimating the three required parameters (frequency, voltage magnitude, and voltage angle). The other input is 220 VAC from power outlet which is used to supply the FDR device with the required power for its electronic circuit operation. Figure 12.4 shows the back panel of the special designed FDR for EWAMS.

Communication Infrastructure for EWAMS

In EWAMS, FDR devices are distributed over a wide area, covering various locations within the boundary of the power system. The FDR devices are then connected to the remote data center through communication network. Public telecommunication infrastructure (Internet) will be used to provide the communication channels between FDRs and the data concentrator server (DCS) in the data center. The existing UMTS mobile communication infrastructure is used to provide Internet access connection for FDRs. High speed packet access (HSPA) USB modems connected to 3G routers are used to provide the Internet access connection.

Client server model is used to provide the communication between FDRs devices and DCS server. Each FDR device is act as a client and requesting to make a connection with the DCS server. The measurements data is transmitted from FDRs devices to the DCS server using a standard network protocols, Transmission Control Protocol (TCP)/Internet Protocol (IP).

4. HELWAN UNIVERSITY HOST SERVERS (HUHS)

The HUHS is data management and processing center operated by several dedicated servers. The logic behind decomposing the HUHS to several numbers of servers is to distribute the computation power which had the advantage of increasing the systems redundancy and reliability. The HUHS consists of four servers connected together through local area Network (LAN). HUHS can divided in terms of functionality to four servers, the data concentrator server (DCS), the real-time application server (RTAS), the Non-Real-time application server (NRTAS), the web server, and the data storage server (DSS). The main functions of each server are mentioned as:

1. Data Concentrator Server (DCS)
2. Real-Time Application Server (RTAS)
3. Non-Real-Time Application Server (NRTAS)
4. Data Storage Server (DSS)
5. Web Application Server (WAS)

5. EWAMS INSTALLATION AND CONFIGURATION

The implemented EWAMS system for the Egyptian power grid consists of 11 FDR units geographically dispersed throughout the boundary of the Egyptian power grid, and a data management system (HUHS) located at, Helwan University. Figure 12.5 shows the location of installed FDRs on the Egyptian power grid.

The installation and configuration of EWAMS consists mainly of two stages. The first stage is the installation and configuration of FDRs and HSPA/3G routers in

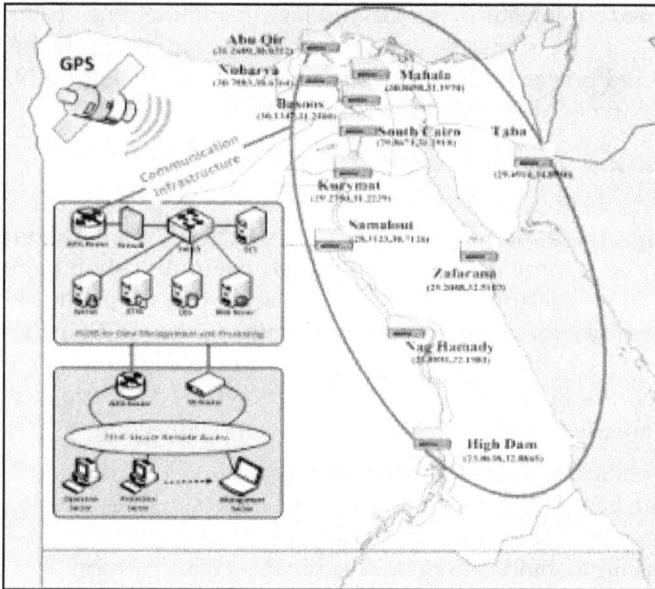

Figure 12.5: FDRs Distribution on Egyptian 500kV- 220kV Grid

the selected power stations. The second stage is the hardware and software installation of the HUHS data management center. In the first stage of installation, the most appropriate VT panel for FDR connection in each power station is initially decided.

Each FDR is configured with the required TCP/IP parameters that enable FDRs to send their measurements data to the designated Data Concentrator Server (DCS). These parameters include the destination IP address or domain name for the DCS and the destination TCP port number. Normally, GPS antenna for FDR is mounted on glass window or on the roof.

Then, each FDR is connected through its Ethernet interface to a 3G router that equipped with HSPA modem. The 3G router is configured to provide the FDR with Internet connectivity for sending the measurements data. The second stage of EWAMS system installation is the implementation of HUHS data center which is consists of two steps. The first step is the Hardware installation of HUHS servers and the second one is the software installation and configuration of EWAMS HUHS data center. Fig. 12.6 shows the four HUHS data center servers installed at Helwan University. For running the EWAMS, a number of software components are installed and linked together on the HUHS servers to provide the different applications. The other EWAMS applications are organized and distributed on the three applications servers "real time server, non-real time server, and web applications server" according to the specificity of each application which are described in more details in the next section.

6. CAPABILITIES OF THE EWAMS

The EWAMS system is now designed. Figure 12.3 shows all the system structure with remote access. The system is capable to cope many of the applications such as

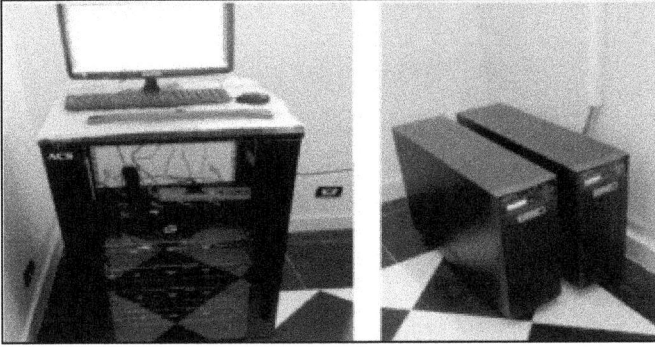

Figure 12.6: HUHS Data Center Servers and UPSs

☆ Wide-area situational awareness

☆ Real-time operations applications

☆ Frequency stability monitoring and trending

☆ Power oscillation monitoring

☆ Voltage monitoring and trending

☆ Alarming and setting system operating limits, event detection and avoidance

☆ Resource integration

☆ State estimation

☆ Dynamic line ratings and congestion management

☆ Outage restoration

☆ Wide area based on protection schemes

☆ Operations planning

Figures 12.7 to Figures 12.11 show some snapshots for the outcomes from the system such as situational awareness and real time application based on EWAMS. However, the EWAMS system introduces huge historical data for post-analysis for fault events, planning and foresting analysis based on frequency and voltage data base.

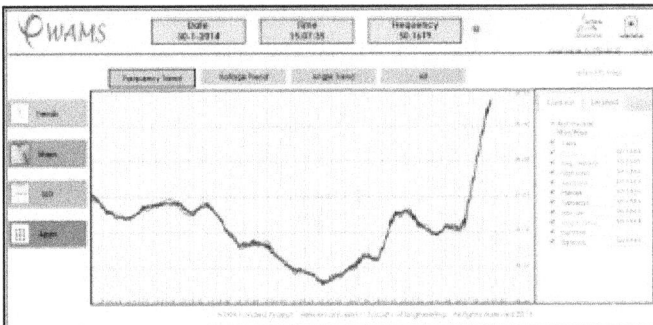

Figure 12.7: Real Time Monitoring on the EWAMS System for the Frequency Trends

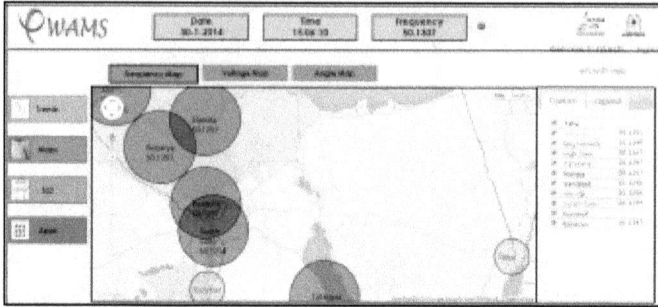

Figure 12.8: Real Time Application on the MAP for the Frequency Monitoring on the Egyptian Grid

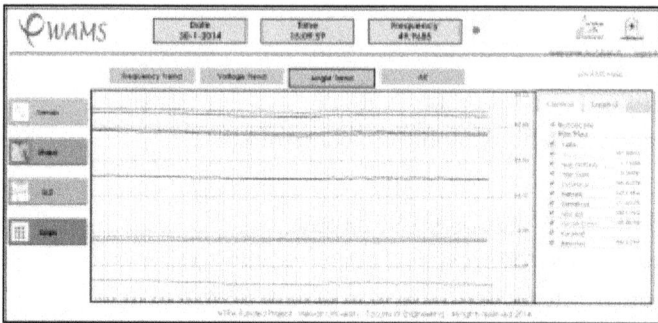

Figure 12.9: Real Time Phase Angle Monitoring Based on EWAMS

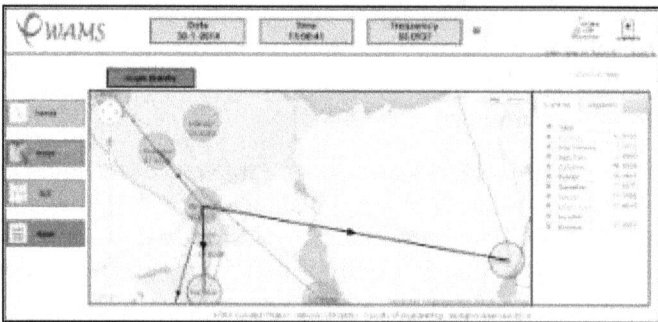

Figure 12.10: Real Time Application for the Stability Monitoring System Based on EWAMS

7. CONCLUSIONS

This project succeeded to monitor the wide-area power system information covering many area on the Egyptian grid using FDRs (family of the Phase Measurement Unit) based on Wide Area monitoring System. This Egyptian Wide Area Monitoring System (EWAMS) momentarily gather data about the system parameters from many area on the Egyptian Grid. The system infrastructure is will explained in the paper. The communication challenges covering many geographically area are also described.

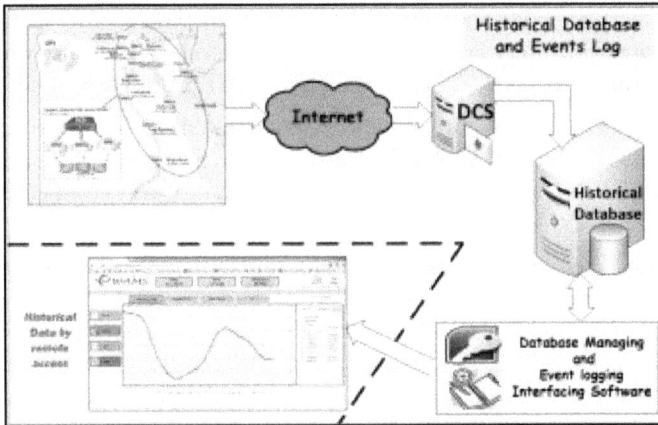

Figure 12.11: Event Log and Historical Measurements Remote Access Procedures

The system remote access avoiding the delay, high band width, single user is also successful achieved in the EWAMS. The system introduced novel application starting from the situational real time awareness in different mode and real time angle stability. The system is able to cover many other novel applications. The system provides Historical Data for post-fault analysis and forecasting process for many years ago.

8. ACKNOWLEDGMENTS

The authors gratefully acknowledge funding and support from National Telecom Regulatory Authority (NTRA), Egypt (http: www.ntra.gov.eg) to implement the network architecture proposed in this work. Also, I would like to thank Egyptian Electricity Transmission Company (EETC) for help in implementing the FDRs devices on real system. More details about the project progress can be obtained on (www.helwan-ntra.com).

REFERENCES

1. M.M. Eissa, Yilu Liu, Mahmoud M. Elmesalawy, and Hossam Gabbar, "Wide Area Synchronized Frequency Measurement System with Secure Communication infrastructure for 500kV/220kV Egyptian Grid", IEEE International Conference on Smart Grid Engineering (SGE'12)-27-29 August, 2012-UOIT, Oshawa, Canada.

2. PSGuarde, Improved Power System Performance through Wide Area Monitoring, Protection and Control, ABB, 2004.

3. A. Bose. Smart transmission grid applications and their supporting infrastructure. IEEE Transcations on Smart Grid, 1(1):11–19, 2010.

4. M. M. Eissa, M. Elshahat Masoud, M. Magdy Mohamed Elanwar" A Novel Back up Wide Area Protection Technique for Power Transmission Grids Using Phasor Measurement Unit", IEEE Transactions on Power Delivery, Vol. 25, No. 1, pp. 270-278, January 2010.

5. M.M. Eissa, T.S.Abdel-Hameed and Adel Sharaf, "Smart Grid with STATCOM Effect Using Harmony Search", IEEE International Conference on Smart Grid Engineering (SGE'12)-27-29 August, 2012- UOIT, Oshawa, Canada.

6. M.M. Eissa, T. S. Abdel-hameed and Hossam Gabbar, "A Novel Approach for Optimum Allocation of Flexible AC Transmission Systems using Harmony Search Technique", IEEE International Conference on Smart Grid Engineering (SGE'13)-28-30 August, 2013- UOIT, Oshawa, Canada.

7. M.M. Eissa, T. S. Abdel-hameed and Hossam Gabbar, "A Novel Approach for Optimum Number and Location of FACTS Devices on IEEE-30 Bus Grid using Meta-Heuristic basd Harmony Search", IEEE International Conference on Smart Grid Engineering (SGE'13)-28-30 August, 2013- UOIT, Oshawa, Canada

8. Miroslav M. Begovic," Smart Grid R&D Perspectives from the US" 3rd IEEE PES ISGT Europe, Berlin, Germany, October 14 -17, 2012.

9. M.D. Hadley; J.B. McBride; T.W. Edgar; L.R. O'Neil; J.D. Johnson, "Securing Wide Area Measurement Systems", U.S. Department of Energy, Office of Electricity Delivery and Energy Reliability, June 2006.

10. Jian Zuo, "The Frequency Monitor Network (FNET) Design and Situation Awareness Algorithm Development", Doctor of Philosophy, Blacksburg, Virginia U.S.A, March 2008.

11. Lang Chen; Markham, P.N.; Yilu Liu, "Wide-area dynamic model validation using FNET measurements," Innovative Smart Grid Technologies (ISGT), 2012 IEEE PES, Vol., No., pp.1,7, 16-20 Jan. 2012.

Chapter 13

Energy Assessment and Implementation of Cleaner Production in a Tea Factory

Warnage Ransisi Keerthi Fonseka[1] and L.I. Palayangoda[2]

[1]*Principal Research Engineer, Industrial Technology Institute,*
Colombo 7, Sri Lanka
[2]*Senior Consultant, Industrial Services Bureau, Kurunegala, Sri Lanka*
E-mail: [1]*keerthi@iti.lk;* [2]*loshanp@isb.lk*

ABSTRACT

The study was conducted to identify losses of energy in a tea factory during the processing steps of tea and to recommend suitable options to improve the condition. The specific study was carried under Switch Asia Project that was funded by European Union in 2009. The factory produced low country grown tea. The factory meets their energy demand from the national grid electricity, diesel and fire wood. The energy management projects proposed in this study falls into three categories such as immediate measures – projects that can be implemented within two weeks, short term measures – projects that can be implemented within 3-6 months and long term measures- projects that is required more than 6 months and also need further study to estimate the savings and/or investments. In this study on cleaner production, options are also categorised as good operating practices, change in technology and training of the staff. Tuning of burners of boilers for optimum combustion, replacing diesel hot air burners with fire wood to minimize fossil fuel usage, replacing conventional type high energy consumed motors with modern types, establishment of fire wood management system for better utilization, operational management to prevent peak loads, introducing preventive maintenance system for accessories and development of staff awareness about energy saving are some options that could save energy considerably. It was revealed that more than 5 per cent electrical energy, 15 per cent fire wood consumption and 15 per cent diesel consumption could be reduced. Hence net carbon dioxide emission to atmosphere could be lowered by 15 per cent.

Keywords: Energy, Efficiency, Assessment, Good operating practices, Cleaner production, Tea.

1. INTRODUCTION

Sri Lanka Tea industry celebrates 145 years commercial history for year 2012. Tea production is one of the main sources of foreign exchange for Sri Lanka and accounts for 2 per cent of GDP and 65 per cent export agricultural revenue to the economy[1]. About 2 million people employed directly and indirectly in the industry and that is about 10 per cent of the population of the country. Tea export earnings USD 1.5 Billion in 2011. The humidity, cool temperatures, and rainfall in the country's central highlands provide a climate that favors the production of high quality tea.

Global tea production reached 4217 million kg and consumption 1719 million kg in 2011. While tea is produced in more than 35 countries, only a handful - China, India, Sri Lanka and Kenya - are responsible for almost three-quarters of production. Almost 56 percent of all tea produced worldwide is consumed locally.

The industry was introduced to the country in 1847 by James Taylor, the British planter who arrived in 1852. Until the 1860s the main crop produced on the island of Sri Lanka, then Ceylon, was coffee. But in 1869, the coffee-rust fungus, *Hemileiavastatrix*, killed the majority of the coffee plants and estate owners had to diversify into other crops in order to avoid total ruin. The owners of Loolecondera Estate had been interested in tea since the late 1850s and in 1866, James Taylor, a recently arrived Scot, was selected to be in charge of the first sowing of tea seeds in 1867, on 19 acres of land[2].

Until 1971, more than 80 per cent of the island's tea estates were owned and managed by British companies. In 1971, the Sri Lankan government introduced a Land Reform Act which gave the state control of the majority of the plantations

Figure 13.1: Tea Estate in Sri Lanka

(which also grow rubber and coconuts for export) leaving about one-third in private hands. Since 1990, a restructuring program has been going on to involve the private sector companies (both Sri Lankan and foreign) as Managing Agents of the state-owned plantations. The long-term aim is for the private managing companies to take on most, if not all, of the financial responsibility and control of the estates, with the government retaining ownership [3].

Table 13.1: Sri Lanka Tea Production–Elevation-wise

Year	High Grown	(per cent)	Mid Grown	(per cent)	Low Grown	(per cent)	Total
2005	80.3	25	55.1	18	181.7	57	317.1
2006	74.7	24	51.6	17	184.5	59	310.8
2007	72.5	24	54.4	17	177.7	59	304.6
2008	84.4	26	49.0	15	185.3	59	318.7
2009	72.8	25	44.7	15	173.1	60	290.6
2010	79.1	24	56.1	17	196.2	59	331.4
2011	79.2	24	52.5	16	196.6	60	328.4

Source: Tea Research Board.

Most of the Ceylon tea gardens are situated at elevations between 3,000 and 8,000 feet in two areas of the southwestern part of the island, to the east of Colombo and in the Galle district on the southern point. In the hot, steamy plains and foothills, the tea bushes flush every seven or eight days and are picked all year round. The finest teas are gathered from late June to the end of August in eastern districts and from the beginning of February to mid-March in the western parts.

Sri Lanka's finest teas are produced mainly from bushes that grow above 4,000 feet. The bushes grow more slowly in the cooler, mistier climate, and are harder to harvest because of the steep angle of the slopes on which they are planted.

There are six main tea-producing areas. Galle, to the south of the island; Ratnapura, about 55 miles east of the capital Colombo; Kandy, the low region near the ancient royal capital; NuwaraEliya, the highest area that produces the finest teas; Dimbula, west of the central mountains; and Uva, located east of Dimbula.

The teas produced in each region have their own individual characteristics of flavor, aroma, and color. Low-grown teas, produced at 1,500 to 1,800 feet, are of good quality and give good color and strength but lack the distinctive flavor and bright fresh taste of the higher-grown teas and are usually used in blending. Mid-grown teas, grown between 1,800 and 3,500 feet, are rich in flavor and give good color. High-grown teas, from heights of between 3,500 and 7,500 feet, are the very best that Sri Lanka produces, giving a beautiful golden liquor and an intense powerful flavor As well as the wonderful black teas, some estates also produce silver tip white tea that gives a very pale straw-colored liquor and should be drunk without milk. All Sri Lanka's black teas are best drunk with a little milk.

Everyday around 300,000 estate workers pluck *several million tea leaves by hand*. This is the first step in the manufacture of quality Ceylon tea. Only the bud and the two youngest leaves are plucked, for it is only these leaves that have the flavor and aroma. In other parts of the world plucking is done by machines. These machines pluck the bud, the young leaf, a lot of *coarse* leaf and few twigs as well. Coarse leaf and twigs just add bulk and not flavor to the tea.

The plucked tea leaf is then brought to the *muster sheds* where they are weighed in, and first quality inspection is made. The leaf is then moved to the factory where they are *withered* using large blowers.

Figure 13.2: Tea Plucking in Sri Lankan Tea Estate

The next step in the manufacturing process involves, cutting the leaves. This brings out the juices and begins the fermentation process. Fermentation is the critical step. The humidity, temperature and fermentation time has to be well controlled or the flavor is lost.

After fermentation is completed, the leaf is *fired*, to lock in the flavor, to dry it and to improve the keeping qualities. Absolutely no preservative or artificial flavoring is added in the manufacture of pure Ceylon tea.

The final step is the separation of the product according to the color and the particle size. Here stringent quality control is done and anything that does not measure up to the standards is rejected.

The finished product is shipped in bulk mainly to Europe, the middle-east, Australia, and North America. Only the best tea is exported.

Present Status

However the fact is that Sri Lanka is no longer the world's largest tea exporter. The industry faces many difficulties with rising cost of production, declining productivity and an acute labor shortage to name a few.

Sri Lanka's Tea production in January to July, 2012 has been reported to be less by 4.4 per cent, from 195.8 mn kg in 2011 to 187.3 mn kg in 2012, during the same period. The total cost per kg of made tea has gone up from Rs. 121.97 per kg in 2001/2002 to Rs. 355.02 per kg in 2010/2011.

We have failed in the search for a sustainable and viable solution to problems the industry is facing. Tea is the most consumed liquid next to water. The authorities should come out with a formula to improve the cost efficiency which is a burden on the industry and productivity of tea production. Reducing the cost of production and increasing productivity alone will not ensure that the Sri Lankan tea industry is able to survive the fiercely competitive global market.

On average the cost of production (COP) for Sri Lanka is the highest in the world. This has affected the country's competitive position in the global market. The primary determinants of cost of production in the tea sector include labor productivity.

Energy Consumption

Sri Lanka produces about 300 million kg of tea. There are about 600 factories in operation in the country. According to an energy audit carried out in the past, the total energy usage for tea processing is about 25.4 MJ per kg of made tea, which consist of about 3.4 MJ of electrical energy and 22.0 MJ of thermal energy. On the basis of 300 million kg of production per annum, the total energy usage by the tea industry is 283 GWh of electrical energy, and 6,600 tera Joules of thermal energy per annum.

The actual energy consumed for tea production is about7.1 kWh/kg. About 85 per cent of this total energy supplied as thermal energy while rest 15 per cent is electricity.

It is estimated that the electrical energy saving potential is as high as 86 million kWh per year or a financial saving of LKR 800 million per year at the current electricity tariff.

At present average specific electrical energy consumption (SEC) of tea sector is 0.63 kWh/kg MT (SLSEA baseline).

Switch Asia Programme

The Food and Beverage industry is a very important sector of Sri Lankan economy that contributes around 10 per cent of the GDP and generate an annual export revenue of US$ 1.4 Bn. However, SMEs in this industry are becoming less competitive due to increase in cost of production due to escalating cost of raw materials, energy, water and low compliance with international standards and safety regulations.

The experts as well as industrialists have equally accepted that switching over to production practices, which ensures sustainable use of natural resources, is the best solution to successfully address these issues. The Promoting Sustainable Consumption and Production (SCP) to Small and Medium Enterprises (SMEs) in the Food and Beverage Industry Project supported by the European Union (EU) would build upon experiences of the past projects and aims to minimize the negative impact of industrial activities on the environmental and society by supporting SMEs in the Food and Beverage Industry to adopt best practices in sustainable consumption and production.Therefore, the main objective of this unique project is to improve the performance of the Food and Beverage sector in Sri Lanka through promotion of cost - effective production techniques and best practices of sustainable production among SMEs.

ORTHODOX TEA : 0.68 kWh/kg made tea

Withering 54%

Grading 13%

Drying 4%

Rolling 29%

CTC TEA : 0.65 kWh/kg made tea

Grading 25%

Withering 15%

Drying 15%

CTC 45%

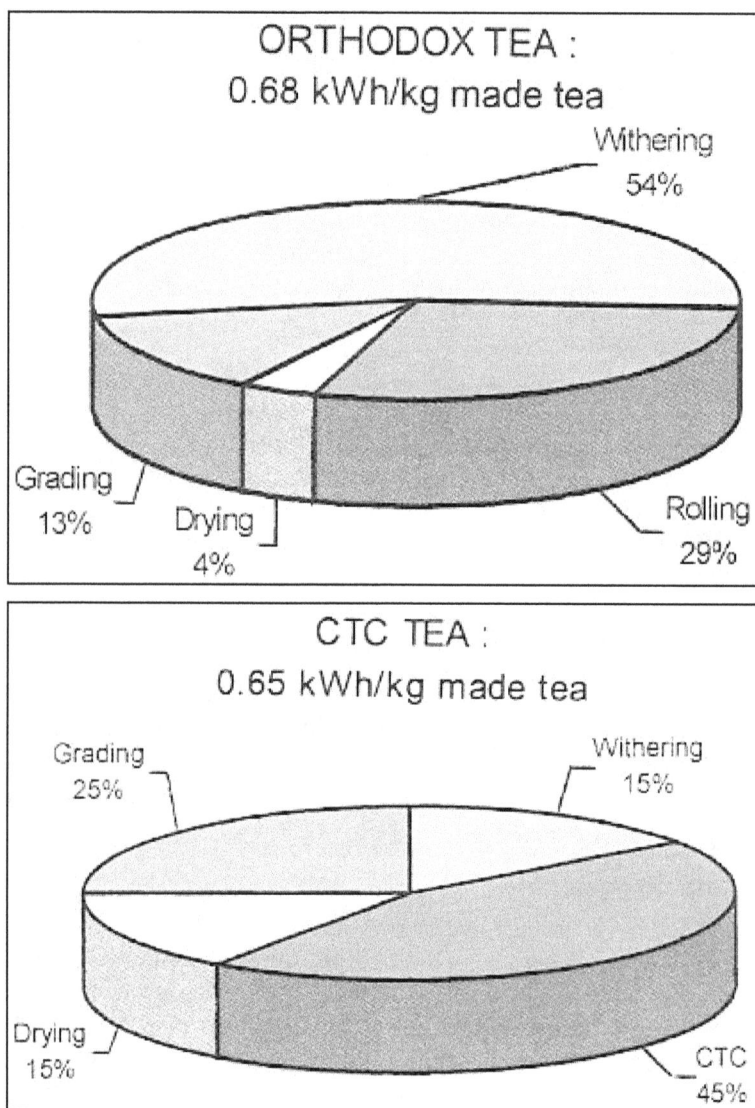

Figure 13.3: Energy Mix of Tea

2. MATERIALS AND METHODS

Cleaner Production *assessment* is one of the specific Cleaner Production diagnostic tools. This is a systematic procedure for the identification and evaluation of Cleaner Production options for the companies that are launching a Cleaner Production project. The methodology allows us to identify areas of inefficient use of resources and poor management of wastes in production.

Planning and Organizing Cleaner Production

The objective of this phase is to obtain commitment to the project, allocate resources and plan the details of the work to come.

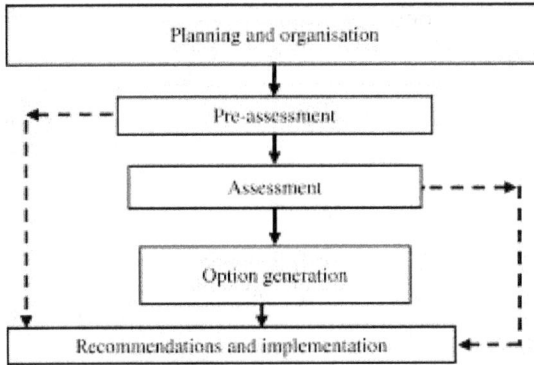

Figure 13.4: Cleaner Production Assessment Methodology (UNEP 1996)

Pre-assessment

The objective of the pre-assessment is to obtain an overview of the production and environmental aspects of a company. Production processes are best represented by a flow chart showing inputs, outputs and environmental problem areas.

Assessment

The aim of the assessment phase is to collect data and evaluate the environmental performance and production efficiency of the company. Data collected about management activities can be used to monitor and control overall process efficiency, set targets and calculate monthly or yearly indicators. Data collected about operational activities can be used to evaluate the performance of a specific process.

Figure 13.5 explains the energy share of tea factory. Thermal energy (via diesel and firewood) is utilized in a dominant way and it is 88 per cent. This is a common situation in most of the Sri Lankan tea factories. The thermal energy load for the

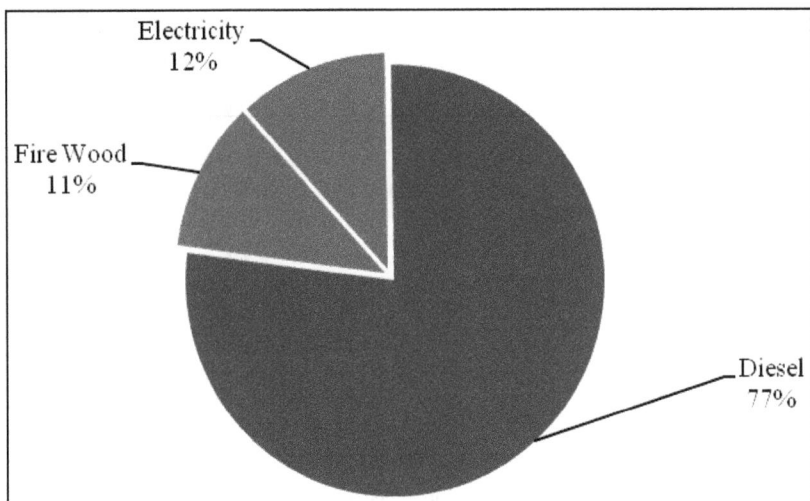

Figure 13.5: Energy Share of a Tea Factory

factory is shared by firewood and diesel for withering and drying processes. Considering the resource wise energy share for the factory and the diesel is the major element (77 per cent), followed by electricity which is 12 per cent. firewood shares 11 per cent of the energy consumption. This relatively high share of Diesel than other tea factories is mainly due to the dryers are operated with diesel burners.

From the analysis from past energy data, Table 13.2 compares the specific energy consumption of the tea factory with that of two national level baselines. The specific electricity consumption of the tea factory is more than 30 per cent over the industry average as per the two sources indicated, thus there is room for improvements. In case of thermal energy (diesel/firewood), there appears to be a higher potential for saving as still the factory is with diesel fired air pre heaters.

Table 13.2: Base Line Analysis

Description	Factory Baseline	SLSEA Baseline for Tea Sector	SWITCH-Asia Baseline for Low Grown Tea Industry
Electrical Energy kWh/kg of Made Tea Energy	0.8372	0.63	0.69
Thermal Energy kWh/kg of Made Tea	6.4471	–	8.05
Fire Wood Utilization kg/kg of Made Tea (MT)	0.2456	1.78 kg/kg of MT	2.44 kg/kg of MT
Diesel Utilization 1/kg of Made Tea	0.5502	–	–
CO_2 emission Tons/Ton of Made Tea	2.0568	–	–

The specific energy consumption for a tea manufacturing facility is an indicator giving of the energy consumption per for a unit of production which in this case is 1 kg of Made Tea. The fluctuation of energy consumption with respect to production is highlighted in the Figure 13.6.

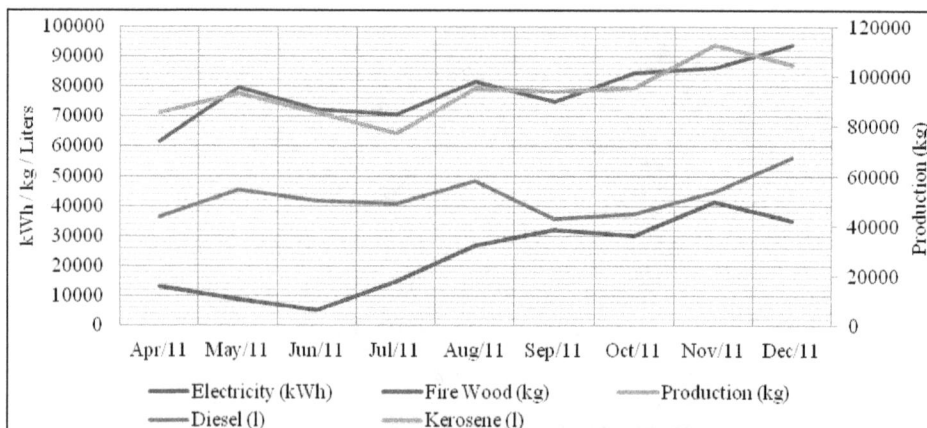

Figure 13.6: Fluctuation of Energy Consumption

Apart from analyzing the past data from energy bills, following equipments have utilized to measure the existing level of electrical energy consumption and the thermal energy consumption of the tea factory:

☆ Dent Data Logger (electrical data collection)

☆ Instantaneous Electrical Power Analyzer

☆ Socket Type Energy Analyzer

☆ Infrared Thermometer

☆ K-Type Thermometer

☆ Digital RH Logger

☆ Flue Gas Analyzer

☆ Anemometer

☆ Lux Meter

Figure 13.7 describes the behavior of the maximum demand which was extracted from the dent data logger fixed at the main panel of the factory. Factory's peak demand is around 280 kVA when operation of all major production units (withering, rolling and drying) takes place simultaneously.

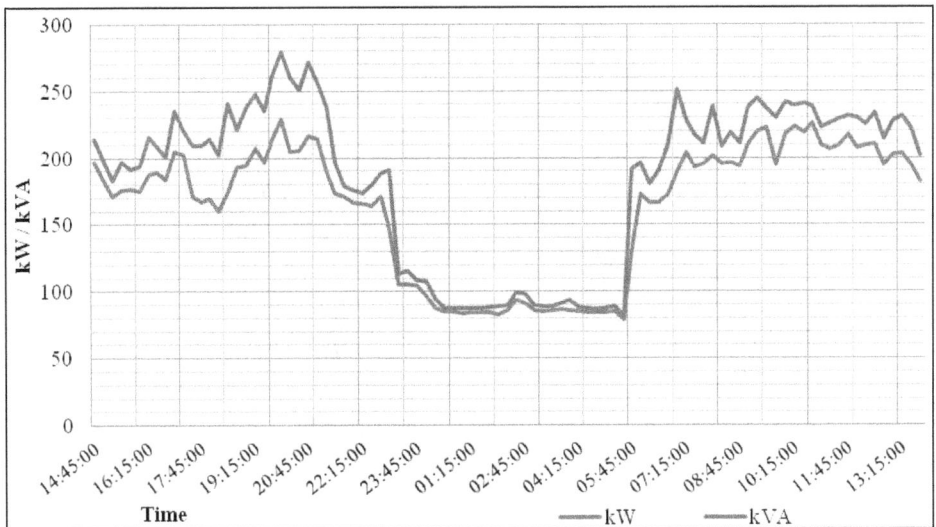

Figure 13.7: Behavior of Maximum Demand

The flue gas analysis for the hot air generator is show in Table 13.3.

Option Generation

The objective of the evaluation and feasibility study phase is to evaluate the proposed Cleaner Production opportunities and to select those suitable for implementation. The opportunities selected during the assessment phase should all be evaluated according to their technical, economic and environmental merits. However, the depth of the study depends on the type of project. Complex projects naturally require more thought than simple projects.

Table 13.3: Flue Gas Analysis

Parameters	Reading
Oxygen (per cent)	16.2
CO (ppm)	558.0
CO_2 (per cent)	4.3
NO (ppm)	28.0
Flue Gas Temperature (ºC)	169.4
Ambient Temperature	26.1
SO_2 (ppm)	44.0
Combustion Efficiency (per cent)	**74.8**

From the measurement taken throughout the day of the assessment following graph depicts the electrical energy share of key sections of the factory. As expected the CTC line 1 and 2 is the most electrical energy consuming area.Recommendations and Implementation

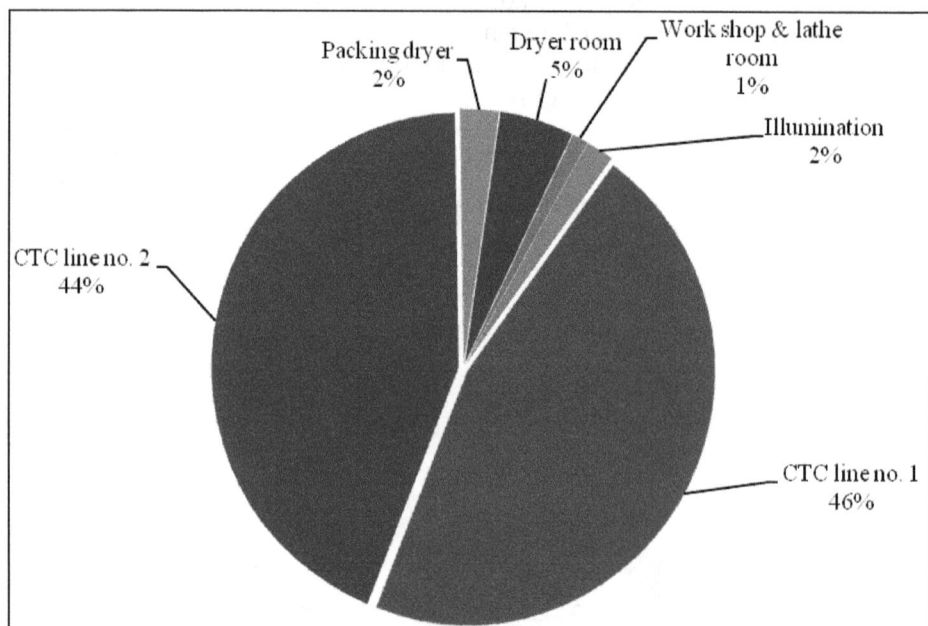

Figure 13.8: Electrical Energy Share

The objective of the last phase of the assessment is to ensure that the selected options are implemented, and that the resulting reductions in resource consumption and waste generation are monitored continuously.

3. RESULTS AND DISCUSSION

It was found that Tea Factory has a great potential of improving the efficiency of energy utilization. It is recommended that low cost measures listed in this study could be immediately implemented in the Factory.

Table 13.4: Monthly Fuel Consumption

Resource	Monthly Average Consumption
Electricity	78,416 kWh; 255 kVA
Diesel	51,529 litres
Fire Wood	23,000 kg

CP category mainly includes Good operating practices (GoP), Change in Technology and Training in the Tea factory. Observations made during the assessment are given below.

Table 13.5: Options Generated during the Assessment

Sl.No.	Observation	Option	CP Category	Cost	Time	Responsibility
1.	Diesel Burners of the hot air generators has not undergone a tune-up	Tune the burners via a flue gas analyzer	GOP	L	I	Factory Maintenance Manager
2.	Hot air generators are operated with diesel	Convert Hot air generators to fire wood	Change in Input Technology	M	I	Factory manager
		or				
		Introduce a steam distribution system via a bio mass boiler to meet the thermal energy demand	Change	H	LT	
3.	Dryers are accom-modated with 40 hp Motors. Measured Data shows that the actual Energy consumption during the operation is 12.8 kW (17.4 hp).	Conduct a further assessment to check the possi-bility to replace 40 hp dryer motors with 20 hp.	Technology Change	H	LT	Factory manager
4.	Absence of fire wood management system	Introduce a shelter for firewood and store the split fire wood	GOP	M	I	Fire store keeper
		Introduce a FIFO system to dispatch the firewood to the boiler	GOP	L	I	
					I	
5.	Work shop activities are carried out during the factory peak time and causing unnece-ssary kVA hike	Aware employees attached to the workshop to avoid unnecessary operations (machining/welding/ etc) during peak time	GOP/ training	L	I	Workshop manager

Contd...

Table 13.5–Contd...

Sl.No.	Observation	Option	CP Category	Cost	Time	Responsibility
6.	Absence of preventive maintenance schedule	Introduce preventive maintenance schedule for key machines				
7.	Lack of employee awareness on energy conservation	Aware employees via posters and SWITCH – Asia presentations Training				

By implementing the above options we expected Tea Factory to reduce energy consumption as follows.

Table 13.6: SWITCH-Asia set targets

Description	Baseline	Expected Reduction
Electrical Energy	0.8372 kWh/kg of Production	5 per cent
Thermal Energy	6.4471 kWh/kg of Production	15 per cent
Fire Wood Utilization	0.2456 kg/kg of Production	
Diesel Utilization	0.5502 l/kg of Production	15 per cent
CO_2 Emission	2.0568 Tons/Ton of Production	8 per cent

4. ACKNOWLEDGEMENTS

Project Director - Swich Asia Project on Sustainable Consumption and Production in Food and Beverage Industry SME sector – Mr. S M Sathiacama, CCC Solutions Pvt. Ltd., Ceylon Chamber of Commerce.

REFERRENCES

1. Annual Report - 2013. Central Bank of Sri Lanka.

2. Energy Management in Tea Sector -2013. Sri Lanka Sustainable Energy Authority.

3. S Sathiacama S M, Palayangoda L, 2012. SWITCH-Asia baseline date for Low Grown Tea Sector".

4. Alternative energy sources and energy saving techniques-2012. Tea Research Institute.

5. Tea Production in Sri Lanka. Wilkipedia, the free Encylopedia.

6. Sri Lanka Tea factories cut energy costs- 2009.- Lanka Business Online.

7. Sustaining Sri Lanka's tea sector is important for the country's future-2013. Sunday Times.

8. Sri Lanka Tea Industry, Tourism and Economy" - H.E. Dr. PalithaKohona, Permanent Representative of Sri Lanka to the United Nations in New York.

9. Jayasingha L.P., Sustainable Indigenous Energy for Tea Industry in Sri Lanka – Bio Energy Association of Sri Lanka.

10. Sri Lanka Tea Industry 2016 and beyond -2014. The Nation.

Chapter 14

Furnace Saves Energy by Oxygen-Enriched Combustion

*Sawaeng Gerdpratoom, Sriwichai Susuk[1]
and Boonchu Leelakajohnjit[2]*

*Thailand Institute of Scientific and Technological Research (TISTR)
35 Mu3 Klong 5, Klong Luang, Pathumthani 12120 Thailand
E-mail: [1]sriwichai_susuk@hotmail.com, [2]boonchu_l@tistr.or.th*

ABSTRACT

Most industry heating process require substantial amounts of energy, which are commonly generated by combusting hydrocarbon fuels. Most combustion processes use air as an oxidant. In many cases, these processes can be enhanced by using the oxidant that contains higher proportion of O_2 than that in the air. Air consists of approximately 21 per cent O_2 and 79 per cent N_2 by volume. Oxygen is used to enrich its ratio in the combustion of melting furnace in several industries. They can increase productivity, reduce overall melt energy cost (save fuel) and emission, while melt loss remain unchanged or slightly decrease. For a certain fuel flow rate, if some furnace parameters are held constant (as time, turbulent level, temperature and product rate), an increas in product rates can be possible with injection of oxygen. This paper presents case of furnaces saving fuel by injection of pure oxygen into air stream mixture (total 24-27 per cent O_2). Pollutant emissions (CO_2, NOx and SOx) also decrease per mass of product.

Keywords: *Melting furnace, Melt energy, Oxygen-enriched combustion, Pollutant emission, Furnace ratio, Particulate matter, Fuel efficiency.*

1. INTRODUCTION

About 90 per cent of the world's energy comes from the combustion of fossil fuels. Energy is needed for transport (land, sea, air), electricity generation, heating in buildings and industrial process (*e.g.* iron, steel, aluminium, paper, cement manufacture). Combustion occurs in boiler, refineries, glass melters, drying kilns, incinerators, industrial oven and is also used to generated energy from biomass (*e.g.* from wood, straw, organic waste). When a fossil such as oil or natural gas is burned, oxygen in the combustion air chemically combines with hydrogen and carbon in the fuel to form water and carbon dioxide, releasing heat in the process. Air is composed of approximately 21 per cent oxygen, 79 per cent nitrogen, and 1 per cent various other gases. Environmental concerns have changed in many industries in the recent past. For instance, strong environmental restrictions have been applied to sulfur oxide emission. As a consequence, high sulfur coal consumption has been significantly reduced in the past few years, due to the high costs of sulfur oxide removal from flue gas. Oxygen enrichment is a well-known technology, used in a variety of industrial applications. Oxygen enriched as well as oxygen-fired combustion system have certain advantages over traditional air-fired technologies, such as lowered emissions, higher combustion and thermal efficiency, higher productivity, improved control etc.

Improving furnace efficiency is a high priority need for metal casting industries. The section of oxygen in the oxidizer stream can be an attractive means of increasing furnace rates of a combustion furnace originally designed to operate with air. For a certain fuel flow rate, if some furnace parameters are held constant (as combustion chamber temperature, turbulence level and time operated) an increase of furnace rates becomes possible with injection of oxygen.

Nitrogen is an inert gas, so it absorbs heat that could otherwise be utilized in the furnace melting. By substituting pure oxygen for a portion of the combustion air, the process will undergo reduced overall gas flow rate and improved thermal efficiency. In most furnace melting processes such a reduction in flue volume will lend itself to more complete melt and, in turn, higher consumption rates.

2. OXYGEN IN COMBUSTION

When air is used as the oxidizer, only O_2 is needed in the combustion process. The N_2 was eliminated from the oxidizer, many benefits. This assumption simplifies the mass and energy balance calculations. The reaction may be written as in (1) and (2) respectively

Mass Balance

Fuel (coke)+Air (21 per cent O_2+79 per cent N_2)+Pure O_2+Pb $\longrightarrow$

Pb+CO_2, H_2O, CO, H_2, SOx, NOx, O_2, ash + heat (1)

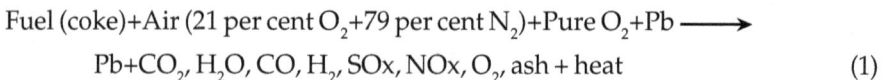

Energy Balance

$$\Delta H_{PR} = \sum_P H_P - \sum_R H_R \tag{2}$$

The amount of energy liberated by the combustion reaction in furnace processes basically follows three different paths:

1. That leaving with the product gases

2. That transferred to the lead, and
3. That lost through the furnace wall.

Oxygen Enriched Air Design

Oxygen has been commonly used to enhance combustion process into the incoming combustion air streams (Figure 14.2). The equivalent overall oxygen in the oxidizer can be calculated as follows:

$$\Omega = \frac{21}{0.21\,(\text{vol}\,\%) + (\text{vol}\,\%\,\text{air})} \tag{3}$$

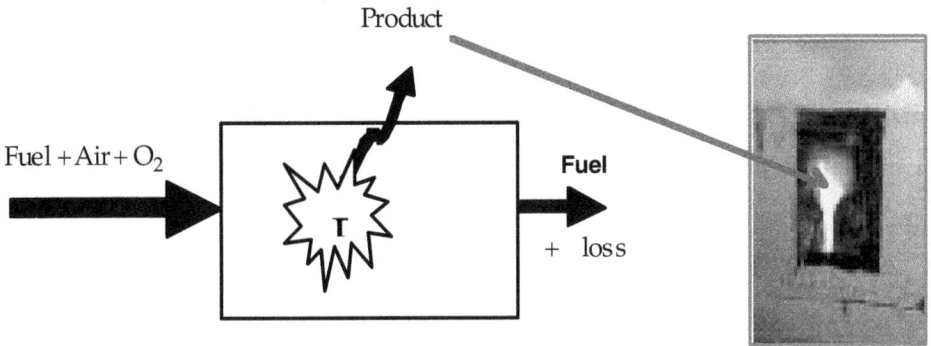

Figure 14.1: A Diagram of Energy Balance

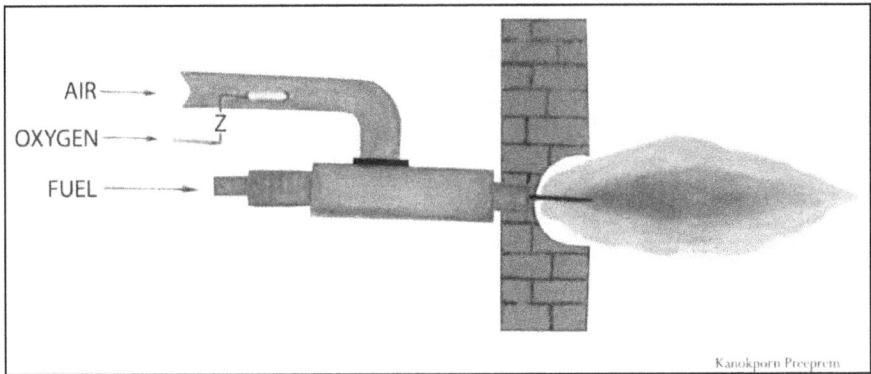

Figure 14.2: An Air/Fuel Process where the Air is Enriched with Pure Oxygen (O$_2$)

3. DEMONSTRATION TEST USING INDUSTRIAL SCALE PLANT

The plant used for the combustion tests consists of lead feeding system, a flue gas treatment system and an oxygen supply 25 per cent and 27 per cent respectively. The tests were conducted in two different modes of operation:

☆ Operation without oxygen enrichment
☆ Operation using oxygen enriched air for primary air

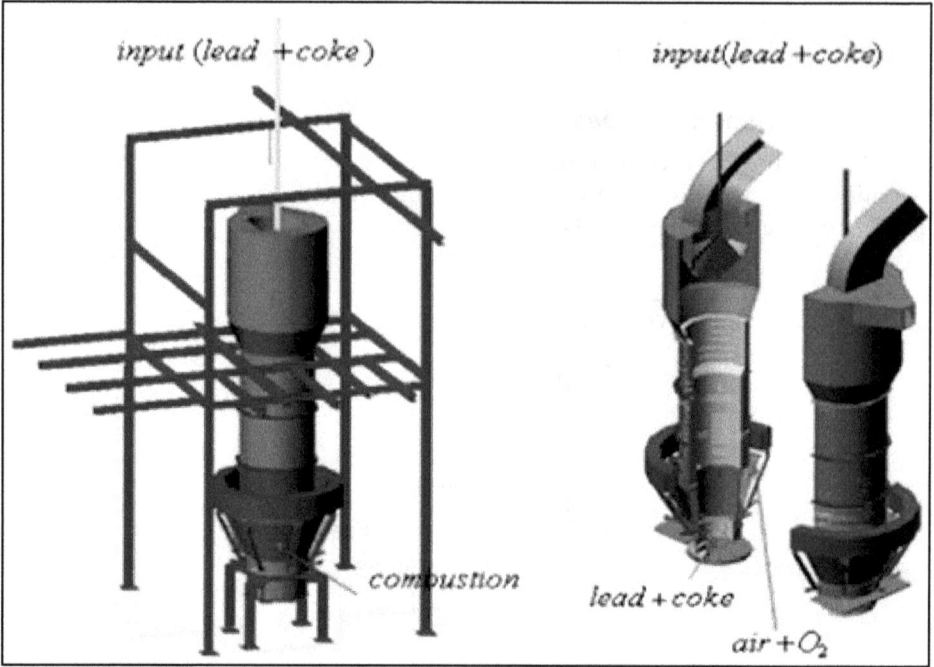

Figure 14.3: Furnace for Melt Lead
(Design by Mr. Sawang Gerdpratoom)

Test Results

Production Increase; The use of oxygen to enrich the combustion air can increase capacity (kg/hr) and furnace can originally be designed to operate with air (Figure 14.4). A sharp increase on the furnace ratio is observed up to 27 per cent of O_2 content in the oxidizer. and also increase in temperature in the combustion area (Figure 14.5).

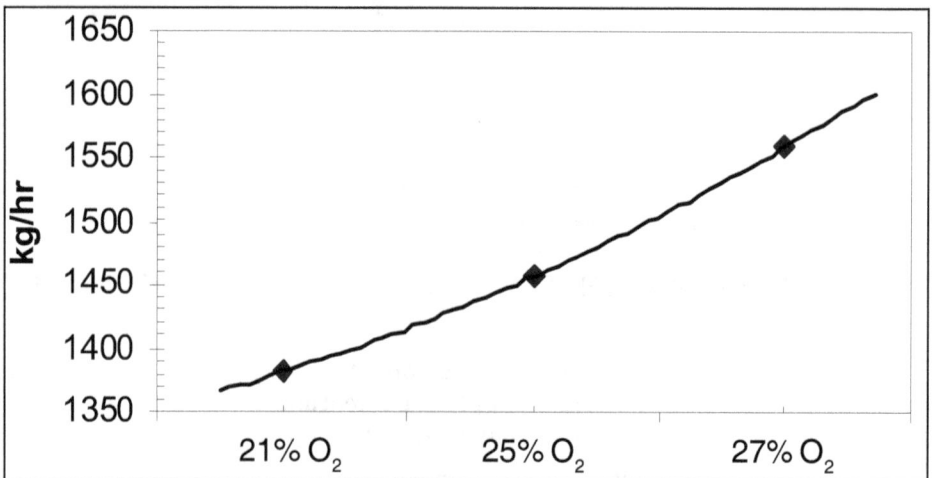

Figure 14.4: Increase Product Lead Melting (kg/hr)

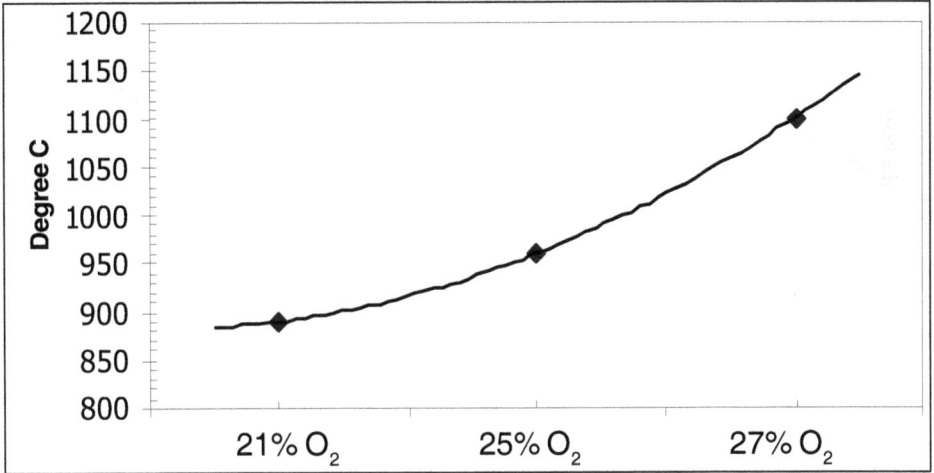

Figure 14.5: Increase Temperature-Melting

Economics and Environment

Operating costs: The two most important concerns of operating cost to be considered for Oxygen Enriched Combustion (OEC) are the fuel and the oxygen. It has been shown that OEC can reduce fuel consumption for a given unit of OEC production (Figure 14.6). There is trade-off between saving in fuel and the cost of the oxygen system. For this test, the primary air was supplied from burner with oxygen concentration 21 per cent.

Furnace ratio is defined as mass flow rate product per mass flow rate fuel, as function of O_2 content in the oxidizer (Figure 14.7). A sharp increase on the furnace ratio is observed up to 27 per cent of O_2 content in the oxidizer. Figure 14.7 suggests

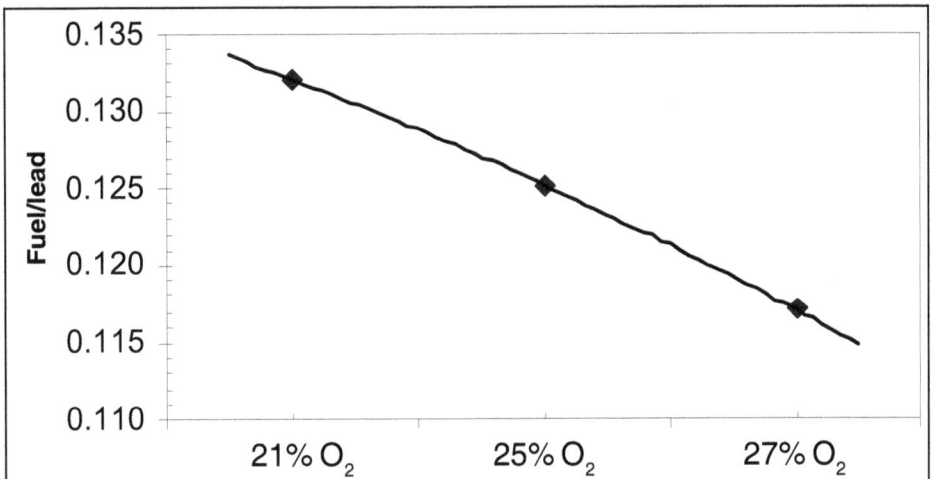

Figure 14.6: Show OEC can Reduce Fuel/Lead

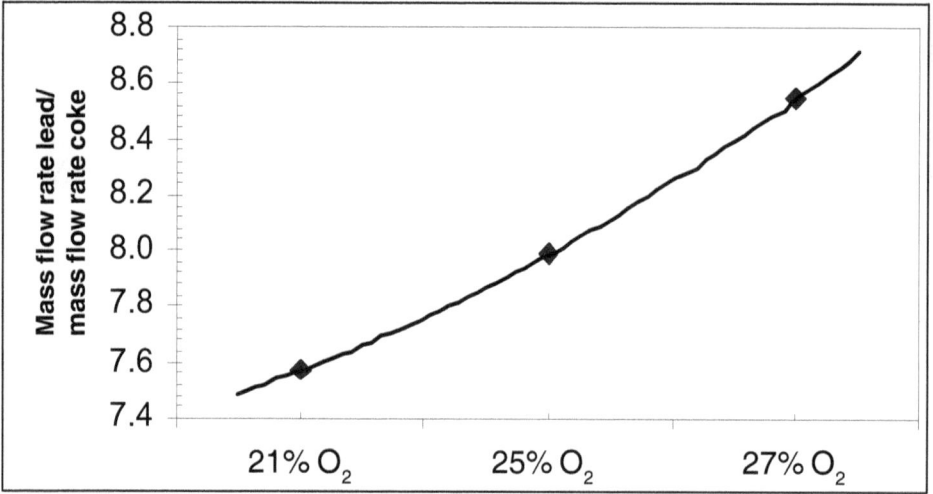

Figure 14.7: Furnace Ratios as Functions of O_2 Content in the Oxidant Gas

that the amount of heat lost through the wall is constant and equal to that for operation with air only underestimates the increase on the furnace ratio with enrichment.

Because the increased fuel efficiency using OEC, the fuel consumption for a given unit of production will be reduced. The cost of oxygen is probably the most variable operating expense. It is very difficult to specify a typical cost. The oxygen cost is 6.81 baht/kg and coke cost is 11.20 baht/kg. Analysising the results of Figure 14.8, it is possible to note that the enrichment in the furnace could reduce the total product cost (kg/baht)/hr.

However; for enrichment above 25 per cent in the furnace, total cost increases with the percent of O_2 increase in the furnace. Enrichment in the furnace is interesting

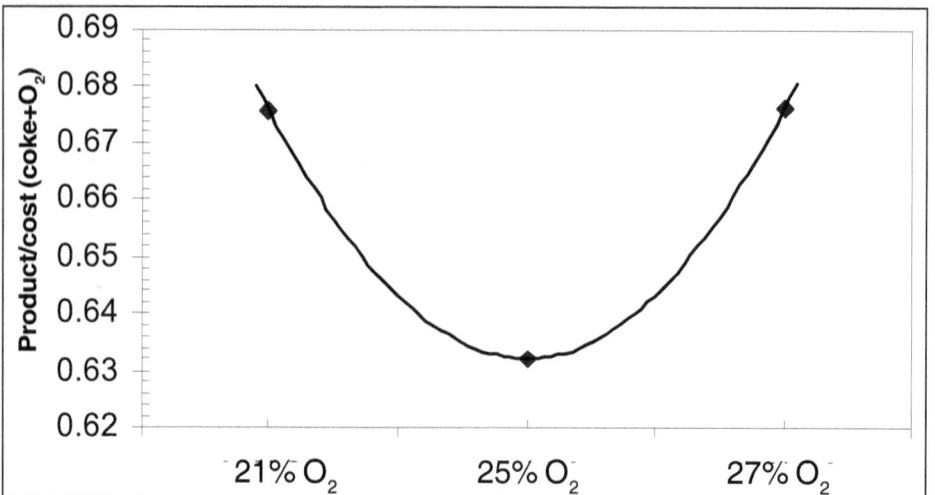

Figure 148: Operating Cost of OEC System.

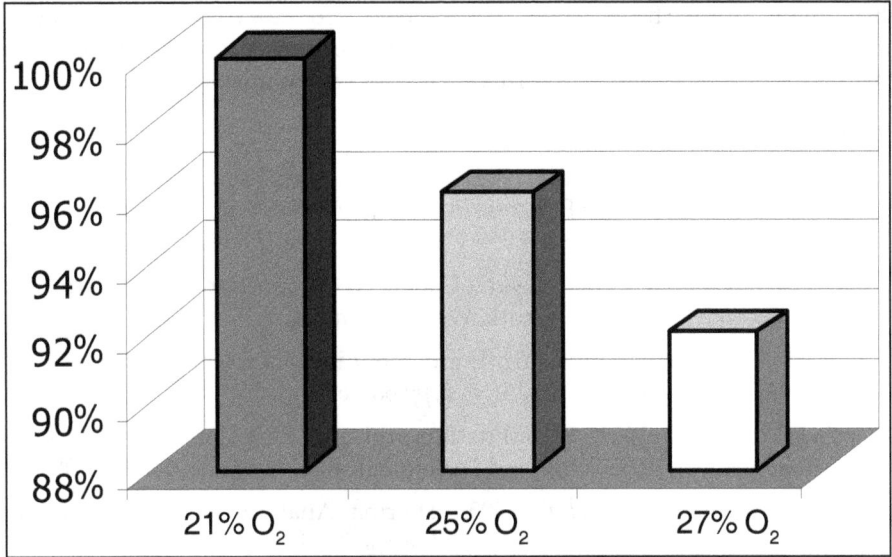

Figure 15.9: Per cent of Particulate Material

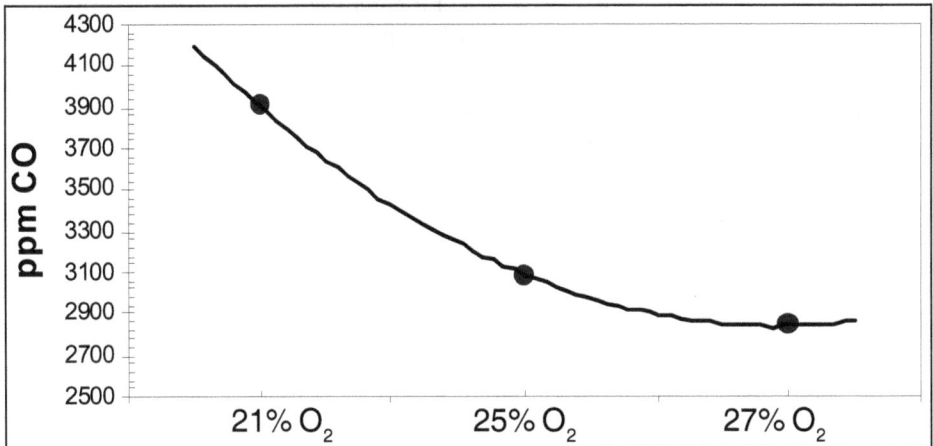

Figure 15.10: ppm of CO

to reduce the amount of emitted particulate material (Figure 14.9). Figure 14.10 presents CO as function of oxygen content in the furnace. Thus, the result of different levels of oxygen enrichment are show in Figure 14.9. The CO formation reduce with oxygen enrichment as seen from Figure 14.10, since the flame temperature reaches higher levels.

4. CONCLUSIONS

This work presents theoretical evaluation of oxygen enrichment combustion applied to the furnace. A simplified mass and energy balance were used to determine the furnace rates. It was shown that a modest increase in oxygen enrichment substantially improve furnace rates. The test results for new combustion technology with oxygen enrichment in this paper. The oxygen-enriched combustion system provides the effective method of furnace with low environmental burdens.

REFFERENCES

1. P.T.Lacava, and A.P. Pimenta. 1999. "Theoretical Analysis of Aqueous Residues Incineration with Oxygen Enriched Flames", *Journal of Heat Mass Transfer, Elsevier Science Ltd Printed in the USA.,1999,Vol.26,No.7, pp. 1029-1040.*

2. Charles E. Baukal Jr. 1998. "Oxygen-Enhanced Combustion". CRC Press., Boca, Raton, Boston, London,New York, Washington,D.C.

3. G.F. Melo. "A Case Study of Air Enrichment in Rotary Klin Incinneration", *J. Comm.Heat Mass Transfer, Vol.25, No.5, pp.681-692.*

4. Sriwichai Susuk. 2007. "Combustion and and Emission in Incinerator ", Thailand Institute of Scientific and Technological Research, September 20th, 2007.

5. Pedro Teixeira Lacava., *et al.,* 2005. "Thermal Analysis of an Enriched Flame Incinerator for Aqueous Residues", *J. Energy, 2005.*

6. Masaharu Kira, Tooru Doi, Shinya Tsuneizumi, Masao Takuma, and Takehiro Kitta. 2001. "Development of New Stoker Incinerator for Municipal Solid Wastes Using Oxygen Enrichment", *Mitsubishi Heavy Industries, Ltd., Technical Review, Vol.38, No.2, Jun. 2001.*

7. Bogdan Horbaniuc., Ovidiu Marin., and Gheorghe Dumitrascu. 2004. "Oxygen-Enriched Combustion in Supercritical Steam Boilers", *J. Energy, Vol. 29, 2004, pp.427-448.*

8. William Bartok. 1989. "Fossil Fuel Combustion a Source Book", *John Wiley and Sons, Inc.*

Chapter 15

Energy Efficiency: Soma City District Heating System Utilizing Waste Heat of Soma B Power Plant

Selçuk Atas[1], Murat Kahraman[2]
and Gökhan Gundogdu[3]

TUBITAK Marmara Research Center Energy Institute, 41470,
Gebze-KOCAELI
E-mail: [1]selcuk.atas@tubitak.gov.tr, [2]murat.kahraman@tubitak.gov.tr,
[3]gokhan.gundogdu@tubitak.gov.tr

ABSTRACT

Lignite fired thermal power plants (TPP) in Turkey were built for only electricity production. However, most of the fuel (2/3rd of total) energy is released through stack flue gases and condenser cooling water in Tpps. These waste heat potential can be used as heat source in district heating system (DH). In a recently completed project in Turkey, fourteen TPPs were analyzed to find out their waste heat potential. Among them, four TPPs were selected and technically analyzed to find out their potential waste heat locations in their thermodynamic water steam cycle. At the end of these analyses, it was decided to utilize Soma B Thermal Power Plant waste heat potential in district heating (DH) system in Soma City as a pilot implementation. In this study, waste heat locations in Soma B TPP thermodynamic cycle is given with their heat potentials by energetic and exergetic analyses. Results of analyses showed that the low pressure turbine inlet stage is most convenient point for steam extraction to meet DH heat demand of Soma City, one of the cities in western side of Turkey with 22,100 dwellings. Some technical conversions to utilize the waste heat are implemented in the TPP and district heating pipes are also installed around the city presently. Currently, the DH system started to supply heat to 900 dwellings. DH system continues to disseminate in Soma and it is assumed that it will reach 8,100 dwellings in 3 years.

Keywords: *Energy efficiency, Exergy, Thermal power plants, District heating, Waste heat, Cogeneration.*

1. INTRODUCTION

Energy can be generated from different sources for different needs. Energy is primarily produced by oil, coal, gas, nuclear, hydro, bio-fuels and renewable sources. Demand sectors can be power (electricity, residential heating/cooling and industries) and transportation. As the sources are limited in the world, energy sustainability is very important for every country in order to manage energy sources in view of future generations' needs. Therefore energy efficiency plays an important role for energy management. Energy efficiency means that less resource will be used for defined and demanded energy needs. Thus, every country should develop their own energy policies in order to use their own resources efficiently, considering total economical and ecological benefits.

Coal-fired thermal power plants (TPP) are most widely used in Turkey. Share of installed power for these TPPs are 22 per cent (totally 12.5 GWe; 14.9 per cent from domestic coal and 6.9 per cent from imported coal) and share of actual electricity production is 28 per cent (15.8 per cent from domestic coal, 12.3 per cent from imported coal) [1]. Having looked at the future of these TPPs in Turkey, the shares for installed capacity from domestic coal sources will be increased by 30 per cent until 2023 [2]. While the electricity produced in these TPPs is only demanded currently, heating/cooling can be also demand side use for houses and industries. The waste heat potential in these TPPs can be used for heating purposes and this kind of application can lead to increase overall efficiency of the plants. Another advantage of utilizing the waste heat of TPPs is that it not only decreases environmental pollution but also increases effectiveness of fossil fuels in a rationalist way. From this point of view, it is most critical to design new TPPs to serve as cogeneration/trigeneration plants and also converting existing TPPs, generating only electricity, as cogeneration/trigeneration plants. It should be noted that before converting these existing TPPs into cogeneration plants, technical and economical feasibility studies should be performed at the beginning.

In the world, there have been some cogeneration conversion projects for TPPs, which was designed to generate only electricity. One example can be seen in the city of Kozani, Greece [3, 4]. In this study, experiences are shared for coal-fired TPP that converted into cogeneration. Heat from TPP is used in district heating (DH) system and at the beginning of the project 17 km transmission line is built between TPP and Kozani City. While 70 MWt heat from TPP was supplied for Kozani City, 30 MWt heat was also supplied to Kozani with peak load boiler plant.

In this paper, conversion for utilization of waste heat for existing Soma B TPP, generating only electricity, is presented. The waste heat will be utilized in DH system in Soma City. In introduction, the TPP and DH system backgrounds are mentioned. Later, thermodynamically energy and exergy analyses for TPP water-steam cycle are presented in order to find correct waste heat sources in the cycle. Then, selection of the waste heat source is set by examining the effect on TPP. Lastly, expected results are presented as a conclusion.

TPP Background

The investigated coal fired TPP in this study is under control of governmental bodies and it was established as sub-critical steam conditions with single reheat unit firing pulverized coal. Full load steam capacity is 540 ton/h at 139 bar superheater outlet pressure with 540 °C superheat temperatures and 448 ton/h from reheater outlet at 31 bar with 540 °C reheated steam temperature (Table 15.1).

Table 15.1: Soma B Thermal Power Plant Design Parameters

Number of Units	6
Unit Power Capacity	165 MW$_e$
Superheated Steam Properties	142 ata, 540 °C, 525 t/h
Reheated Steam Properties	32 ata, 540 °C, 448 t/h
Steam Pressure and Temperature @ Condenser Outlet	0.07 ata, 38 °C
Condenser Cooling Water Temperature and Flow Rate	27 °C, 22 600 t/h
Boiler Feed Water Inlet Temperature	234 °C
Fuel Type	Soma Lignite
Fuel LHV for Unit 1-4	2 400 kcal/kg
Fuel LHV for Unit 5-6	1 550 kcal/kg

The TPP, under investigation, has six units with 165 MWe each, single pass with natural circulation, drum type boiler firing lignite and furnished with six high speed impact mills. The boiler was designed with lower calorific value of 1550 kcal/kg, 18.8 per cent moisture, 64.2 per cent ash, 15-20 per cent volatile matter, 1.5 per cent sulfur and 17-25 per cent carbon contents.

Continuous mass flow diagram for water steam cycle is shown in Figure 15.1, that is created by employing design parameters and in-situ investigations. Flow diagrams are valid for all in 6 units in Soma B TPP because all units are similar to each other.

Soma City and DH System

Soma is a city and district of Manisa Province in Aegean region of Turkey. According to the 2009 census, population of the district is 101,011 of which 74,158 live in the city of Soma. The district covers an area of 839 km^2 and city lies at an elevation of 161 m (Figure 15.2). In the City of Soma, buildings are heated by coal based heating systems. 70 per cent of dwellings are heated by old fashioned stoves and 30 per cent of dwellings are heated by central heating boilers in the buildings which have own radiator systems. There are 22,100 dwellings in Soma City center. At the beginning it is planned to install a DH system for 8,100 dwellings in 3 years.

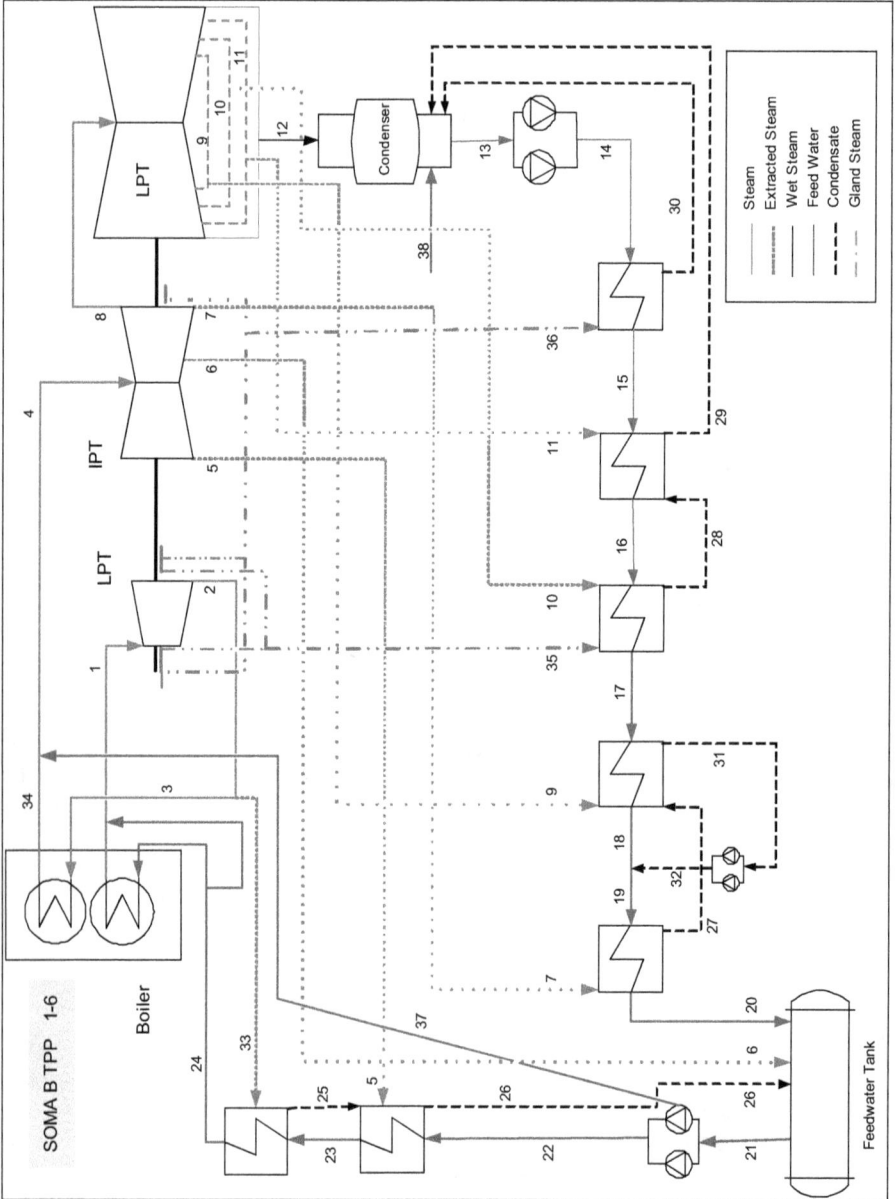

Figure 15.1: Continuous Mass Flow Diagram of Soma B TPP

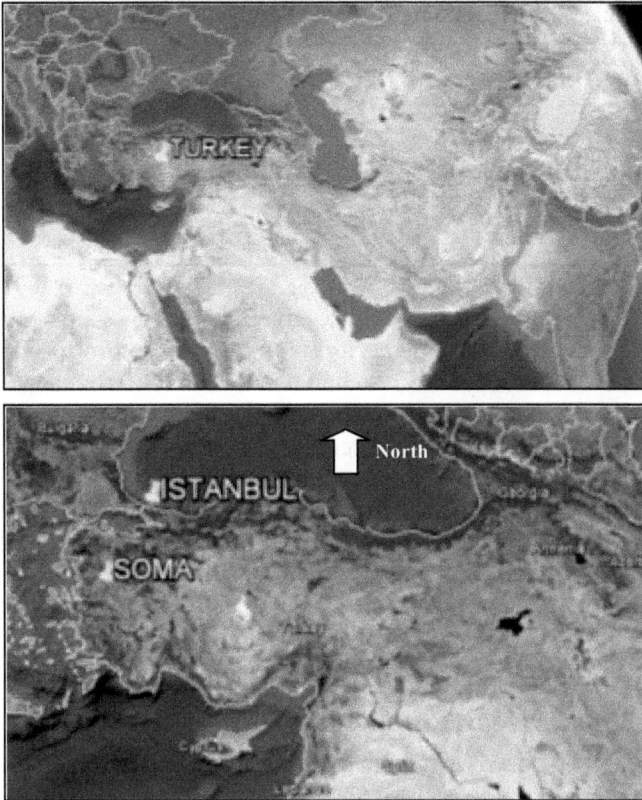

Figure 15.2: Location of Soma City

In Soma, winters are cold and the rainiest season, whereas summers are the least rainy season. July is the hottest month and the average temperature of Soma is 23-24 °C. In winter, Soma is affected by cold air streams coming from Balkans and average temperature is low. January is the coldest month with an average temperature of 3-5 °C. Typically the weather is snowy in the region.

In order to determine the DH system heating capacity and demand from TPP, the heat load of a 100 m² dwelling was calculated by using hourly temperature and solar radiation distribution in a year. The calculations are performed with the standard of TS825 and it is assumed that there will be a heating at an outdoor temperature below 15°C. The lowest ambient temperature in the calculation with respect to the location of Soma City is 1.9 °C. As a result, for one dwelling, it was found out that the peak heating demand is 7.2 kWt and domestic hot water heat demand is 1.9 kWt. Total heating energy demand for one dwelling in a year was determined as 17.1 MWth. For 8,100 dwellings, the peak heat load is 64 MWt and total energy demand from TPP with heat losses was 153,200 MWt/year (Table 15.2) [5].

Table 15.2: Heat Load for Soma City (For 8,100 Dwelling)

Dwelling		8,100
Peak Heat Load	MWt	64
Yearly Heating Energy Load	MWt/year	128,500
Yearly Hot Water Energy Load	MWt/year	10,200
Heat Losses	MWt/year	14,500
Total Heat Load	MWt/year	153,200

A typical DH system is formed by three main parts. These are heat production, transmission and distribution sides (Figure 15.3). Heat is produced in TPP and transferred to the DH water by a heat exchanger (1). Condensed steam (2) in TPP can be fed into the TPP cycle from the feedwater tank by a variable speed controlled condense pumps. At DH side, return circulating water comes into the TPP DH heat exchanger (1) and heated up to desired heating temperature then it is pumped (3) to the city by transmission pipeline. The heated circulating water then flows into dwelling heat exchangers (4) and gives its energy to the dwelling circuiting water (5). The dwellings are heated by that water with help of radiators. Also, hot tap water is obtained from hot DH water inside the dwellings. Heat used in the dwelling and total heat for DH system is measured by heat meters for billing operations (6). The leaked water in circulating water is compensated as well (7).

Analysis of Waste Heat Potentials

Having established the city heat demand in the introduction, in this chapter, it will be described how to explore the waste heat in the TPP water steam cycle. In the content of this study, water steam cycle of Soma B TPP was thermodynamically (energetic and exergetic) analyzed. The analyses have been performed by specially developed comprehensive energetic model. In the literature, it is seen that researchers [6-8] tried to evaluate the performance of the TPPs with thermodynamic first (energy) and second (exergy) laws. While the first law provided them a quantitative interpretation of the thermodynamic analysis with energy balance, second law gave qualitative information with exergetic balance. In other words, first law describes the energy potential of the sources but it does not mention its quality whether it can be used for the application or not. On the other hand, exergy gives the quality of the source and what one can sacrifice if that source is used in the application.

In order to find heat extraction point in the TPP water-steam cycle (Figure 15.4) and perform energetic and exergetic analyses, some assumptions have been done for calculations as follows:

☆ Kinetic and potential energy changes are neglected in the energetic and exergetic calculations

☆ Dead state temperature and pressure are assumed as 15°C and 1 bar

☆ The coal ultimate analysis is assumed as 20.8 per cent carbon, 1.8 per cent hydrogen, 0.45 per cent nitrogen, 1.0 per cent sulfur, 13.2 per cent oxygen, 44 per cent ash and 18.8 per cent moisture

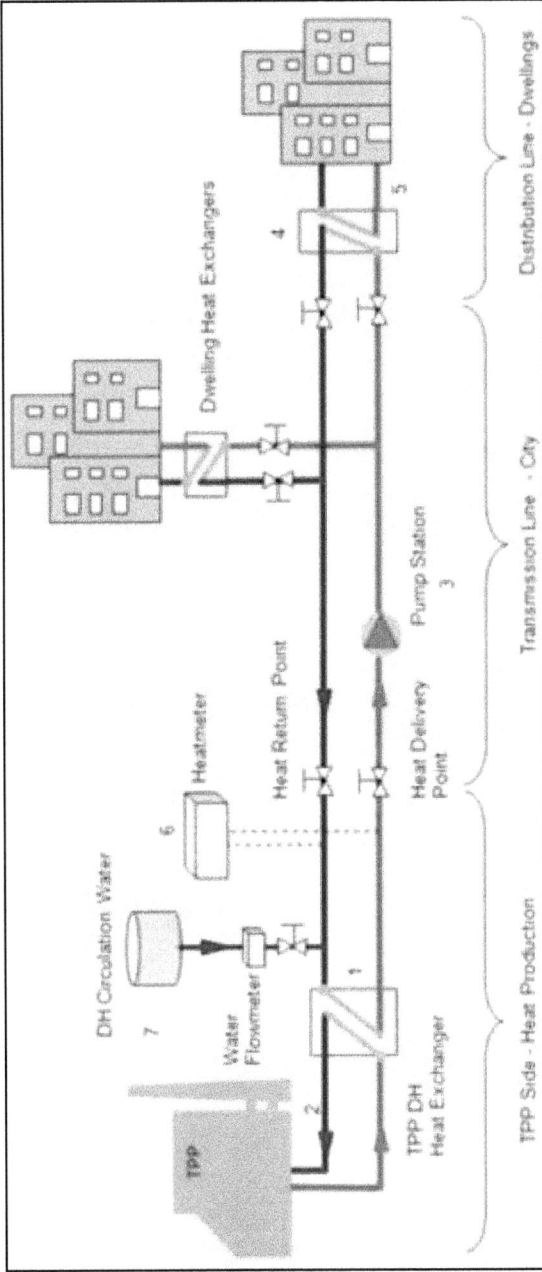

Figure 15.3: Typical DH System

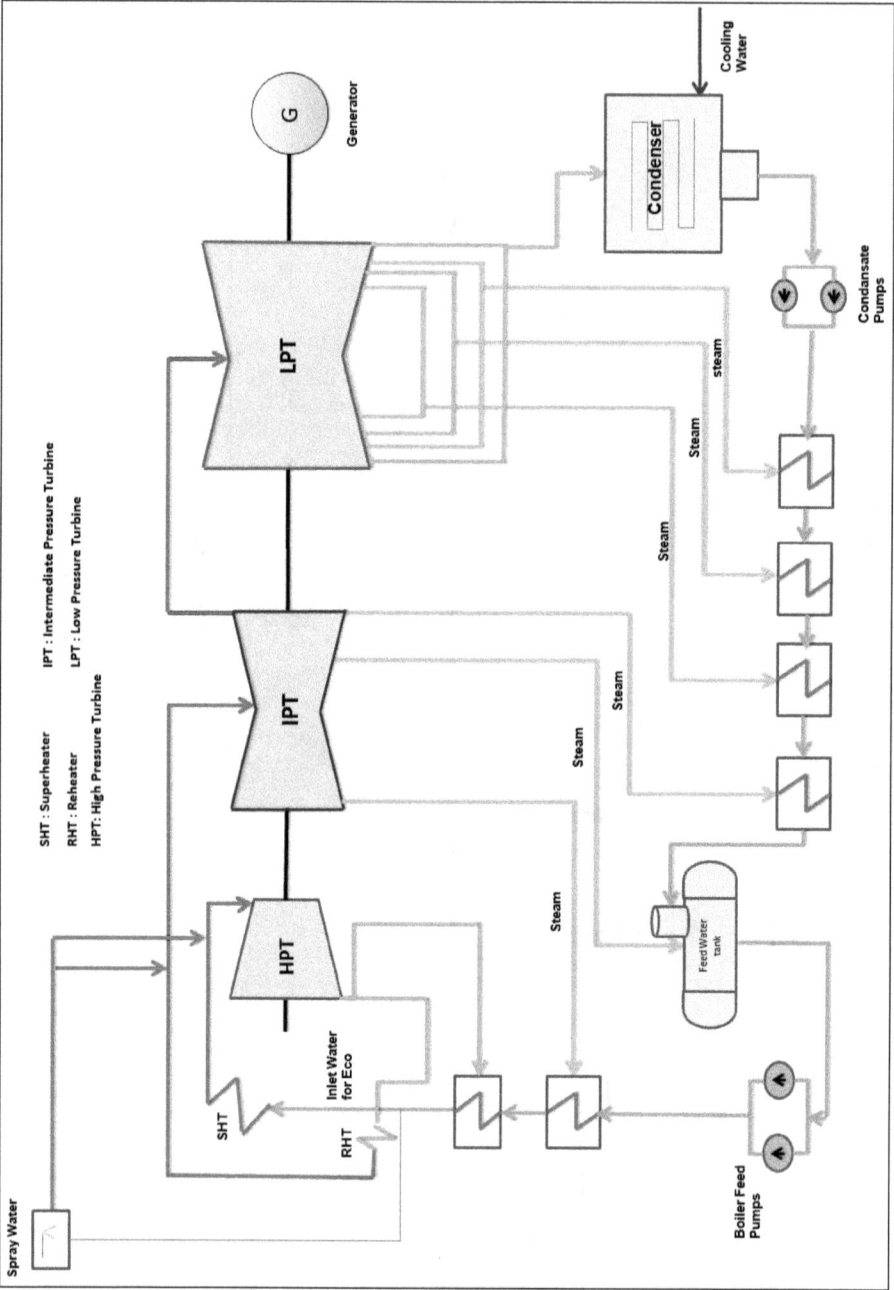

Figure 15.4: Diagram of Soma B Thermal Power Plant Simulation Model

✰ All coal ultimate analyses and fluid properties, turbine, boiler and pump efficiencies are provided from TPP design documents

For thermodynamic analyses, TPP water steam cycle model (Figure 15.4) was developed in plant simulation software based on a constant mass flow diagram which has been used in potential analysis of the plant. Results of developed model were validated with design parameters of the plant. Predicted and design values are in good agreement such that maximum difference is about per cent 3. Furthermore simulations were carried out at partial load operating conditions and validated with guaranteed values.

According to thermodynamic calculations, waste heat and other heat potentials are analyzed by the help of energy and exergy of the heat potentials. The energetic and exergetic analyses for waste heat potentials are performed for condenser, stack gases, feed water heaters and low pressure turbine inlet.

Condenser

Waste heat potential of condenser, energy and exergy of waste heat has been determined for condenser by energy and exergy equations 1 and 2. It is assumed that latent heat of wet steam is potential heat source.

$$Q_{con} = (m_s h_s)_{in} - (m_s h_s)_{out} = (m_{cw} h_{cw})_{in} - (m_{cw} h_{cw})_{out} \qquad [1]$$

$$Ex_{con} = m_s[(h_s - h_0) - T_0 \cdot (s_s - s_0)] \qquad [2]$$

Here m_s, m_{cw} are the mass flow rate; h_s, h_{cw} are the enthalpy values of the condenser steam, cooling water at the inlets and outlets. s_s is entropy of the steam at the inlet of the condenser. T_0, h_0 and s_0 are dead state temperature, enthalpy and entropy at 15°C respectively.

Stack Gases

Flue gases are extracted from TPP stack at a temperature approximately 157 °C by the help of induced draft fan. Due to the sulfur content of the coal, the flue gas contains some content of SO_2. If flue gas temperature decreases the condensation temperature, then the acidic liquid is formed and it may harm the equipments. Thus, it is dangerous to decrease the temperature below 100°C. Thus, the waste heat inside the flue gas is utilized until that condensation temperature limit and its potential is calculated by the equation 3. The exergy of the gas is calculated by equation 4.

$$Q_{fg} = (m_{fg} h_{fg})_{in} - (m_{fg} h_{fg})_{out} \qquad [3]$$

$$Ex_{fg} = m_{fg} C_{fg} \left[(T_{fg} - T_0) - T_0 \cdot \ln\left(\frac{T_{fg}}{T_0}\right) \right] \qquad [4]$$

Here T_{fg}, m_{fg}, h_{fg} and C_{fg} are temperature, mass flow rate, enthalpy and specific heat of flue gas respectively. T_0 is dead state temperature.

Feed Water Heaters

The steam from the turbines for feed water can be used for district heating applications, if it is technically and economically feasible and it can be utilized in a reliable way. These kinds of applications were employed in some TPPs [8-10]. As heat is not needed in summer time, then it is conveyed into feed water heating again. In calculations, heat potential of extracted steam calculated up to latent heat of condensation at 70°C. In the system, there are four low pressure (LP FWH) and two high pressure (HP FWH) feed water heaters. Energy and exergy values for the heaters can be calculated by the equations 5 and 6 below.

$$Q_{heater} = (m_s h_s)_{in} - (m_s h_w)_{out}$$ [5]

$$Ex_{heater} = m_s[(h_s - h_0) - T_0 \cdot (s_s - s_0)]$$ [6]

Here m_s is the steam mass flow rate; h_s, h_w are the enthalpy values of the heater steam and water at 70°C. s_s is entropy of the steam at the inlet of the heater. T_0, h_0 and s_0 are dead state temperature, enthalpy and entropy at 15°C.

Low Pressure Turbine Inlet Steam

Apart from these potential heat sources, inlet steam for the low pressure turbine was examined both for energetic and exergetic point of view. It is assumed that just 20 per cent of steam can be extracted for utilization in order not to alter turbine operation. This kind of applications can be seen at some TPP like in Deva TPP in Romania and Narva TPP in Letonia (Figure 15.5). The heat potential was calculated until the extracted steam was converted into full saturation and up to heat of condensation at 70°C. This potential can be calculated with energy and exergy balances by the equations 7 and 8 below.

$$Q_{lpt} = (m_s h_s)_{in} - (m_s h_w)_{out}$$ [7]

$$Ex_{lpt} = m_s[(h_s - h_0) - T_0 \cdot (s_s - s_0)]$$ [8]

2. RESULTS AND DISCUSSIONS

According to the energetic and exergetic calculations, heat potentials and their exergies are given in the Table 15.3. The evaluations are carried out in the Table 15.4.

After validation of the developed simulation model and thermodynamical calculations for waste heat potentials, in order to determine waste heat point in the plant cycle, some technical aspects are drawn as follows [11,12].

1. Specified heat sources should not be in liquid media: This requires too much liquid to utilize in DH system as latent heat in steam is not used.

2. Extracted steam temperature should be convenient for DH application: Generally 70-100 °C are convenient for radiator heaters at homes. In order not to lose more exergy by heat transfer of different temperature media, the temperature levels should be compatible both for potential heat source and DH water side.

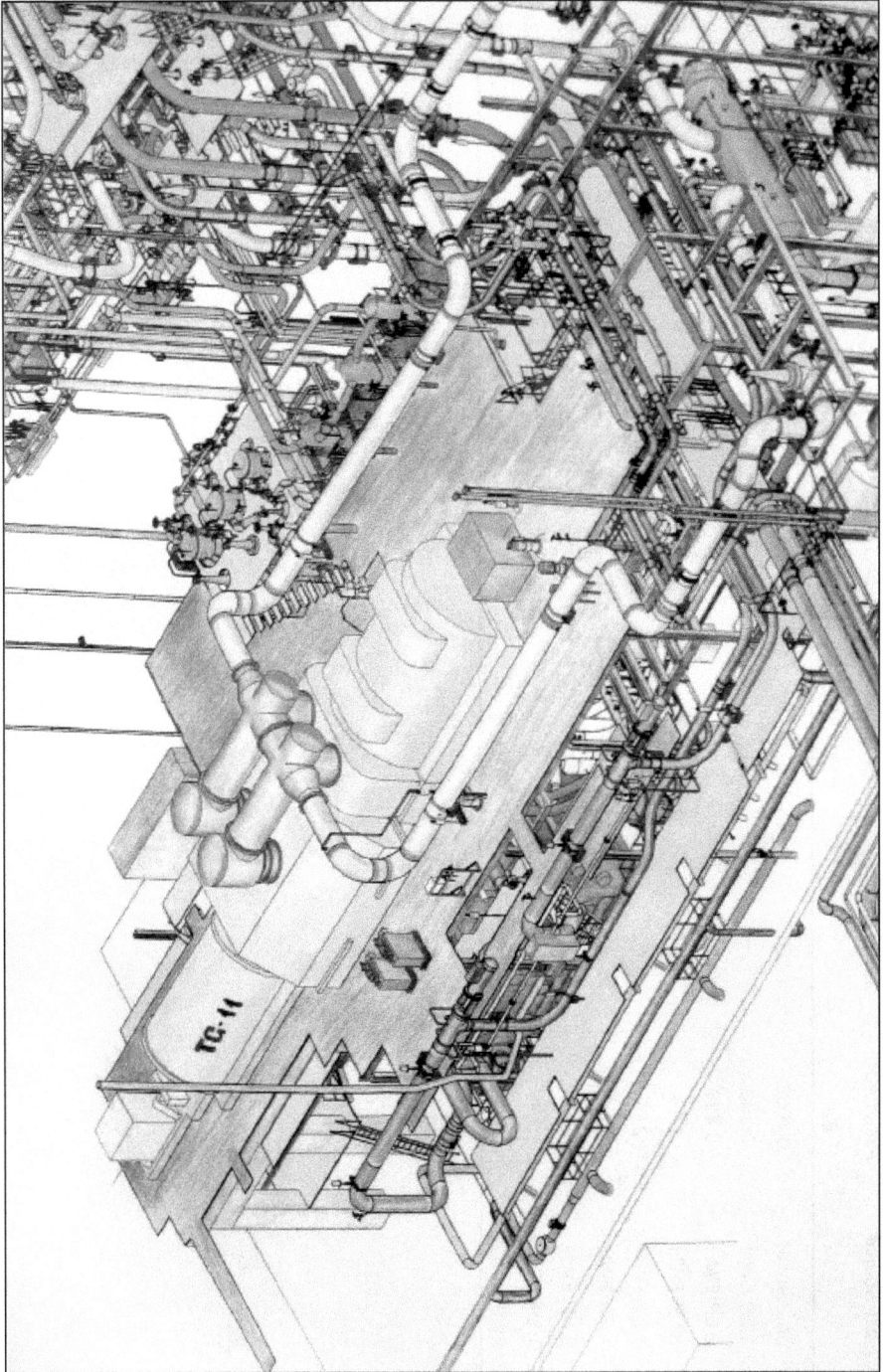

Figure 15.5: Narva TPP DH System with Extracted Steam before Low Pressure Turbine

Table 15.3: Heat Potentials of Soma B TPP

Heat Source	Pressure Bar	Inlet Temp. °C	Outlet Temp. °C	Saturation Temp. °C	Energy Rate kW	Exergy Rate kW	Exergy kJ/kg-K
Condenser	0.0068	38.6	38.6	38.6	209,487	16,425	177
1.LP FWH	0.196	60	59.6	60	9,959	1,400	323
2.LP FWH	0.496	91	81.1	81	9,169	1,350	462
3.LP FWH	1.16	161	70	104	10,366	2,536	613
4.LP FWH	2.29	225	70	124	11,018	3,142	750
1.HP FWH	12.2	420	70	189	14,044	5,434	1,164
2.HP FWH	29.8	343	70	234	39,234	16,299	1,164
LPT Inlet Steam	2.3	226	70	124	50,641	49,367	462
Stack Flue Gas	–	157	100	–	16,854	8,105	27.8

Table 15.4: Evaluation of Heat Potentials

Waste Heat Source	Waste Heat Potential (MW$_t$)	Source Temperature (°C)	Advantages for Utilization in DH	Disadvantages for Utilization in DH
Condenser	209.5	38.6 Steam	Highest potential exists (Figure 14.6). Economical since low exergetic heating potential is used (Figure 14.7). There is no effect on electric generation amount with second lowest exergy value (Figure 14.7).	Very low temperature for radiators (Figure 14.6). Radiators are generally operated at temperatures 50-90°C. DH can use that source by using heat pump or wall/ground heating. Its temperature compatible to fish feeding and green houses. High investment and operational costs for DH as high flow rates are required to maintain DH demand.
Stack flue gas	16.8	157 Gas	Fully waste heat emitted to the atmosphere is utilized with lowest exergy (Figure 14.7). There is no effect on electric generation amount (Figure 14.7).	Stack suction decreases so there can be a need to increase fan power or use booster fan. High initial investment cost as gas has a low heat transfer coefficient. Limited waste heat potential compared to Soma City heat demand (Table 14.2). It is dangerous for TPP if H_2SO_4 is formed due to the decrease in saturation temperature. It should not be reduced below saturation temperature.
Feed water heaters		Steam	1. and 2. Low pressure heaters have low source temperatures and exergies. Thus, their unit heat costs are low (Figures 14.6 and 14.7).	Feed water tank temperature depends on the feed water heaters. Extracted steam from feed water heaters for DH should not reduce the feed water temperature to its limit temperature value. As the TPP is not designed for DH system, point for extracted steams is not determined in the design period. Thus, sources are limited in potential. As the extracted steam gets close to the boiler, the exergy of the source increases. This causes costs of the heat potentials. (3.4. low pressure heaters and 1.2.3 high pressure heaters).
1. LP FWH	9.9	60		
2. LP FWH	9.2	91		
3. LP FWH	10.4	161		
4. LP FWH	11.0	225		
1. HP FWH	14.0	420		
2. HP FWH	39.2	343		

Contd...

Table 15.4–Contd...

Waste Heat Source	Waste Heat Potential (MW$_t$)	Source Temperature (°C)	Advantages for Utilization in DH	Disadvantages for Utilization in DH
Low pressure turbine inlet	50.6	226 Steam	Not so much work is required while adapting the DH system. There is little change in electric generation amount when compared to 3.4. low pressure heaters and 1.2.3 high pressure heaters (Figure 14.7).	Extracted steam before low pressure turbine causes some decrease in pressure and it increases the loading in turbine blades. Thus the steam extraction is limited upto 20 per cent of total steam amount before low pressure turbine. Limited but significant heat potential exists. This is for one unit and there are 6 units in TPP and totally 303.6 MWt. That is enough for all Soma City DH system.

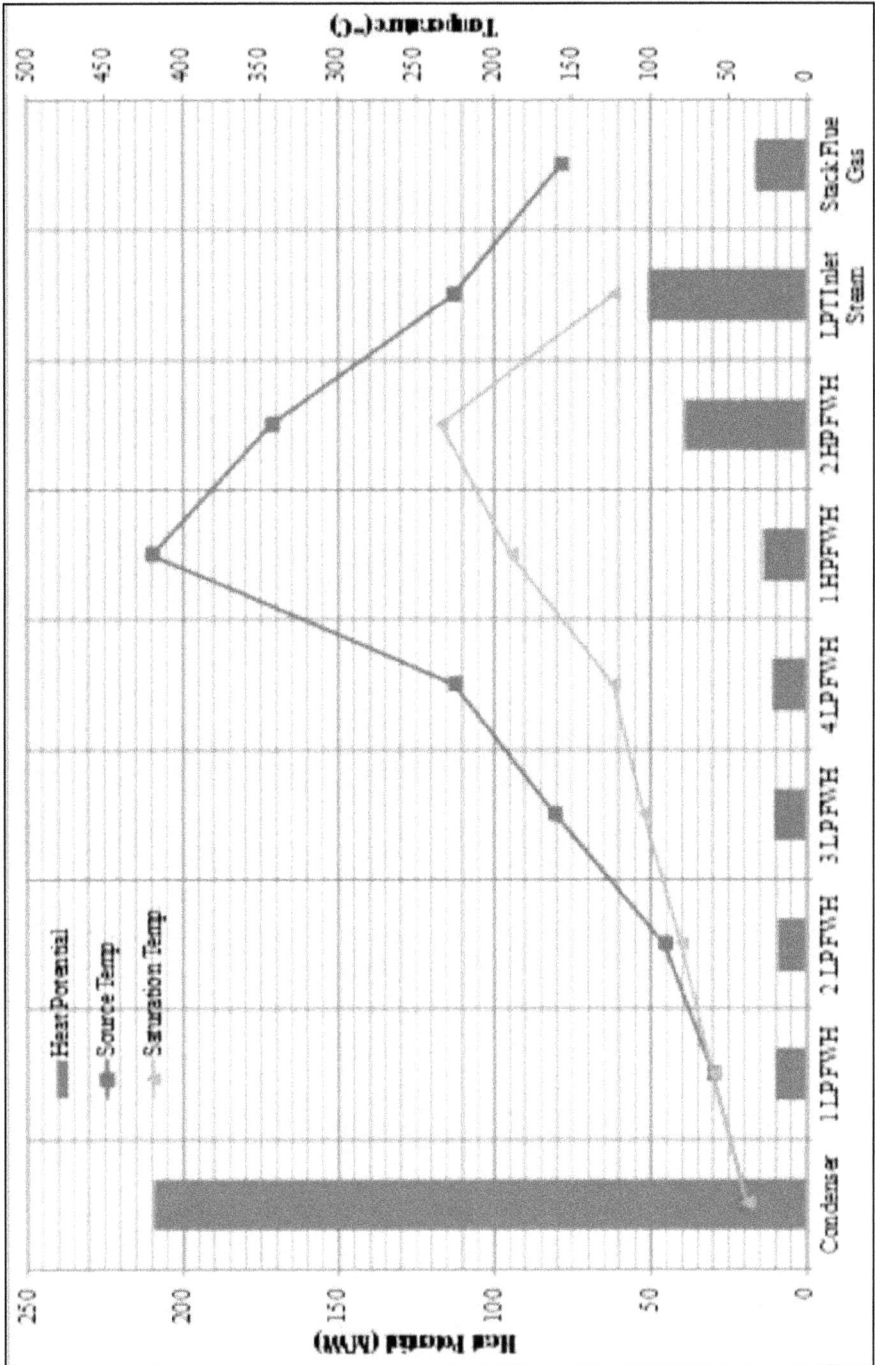

Figure 15.6: TPP Heat Energy Potentials with Source and Saturation Temperatures

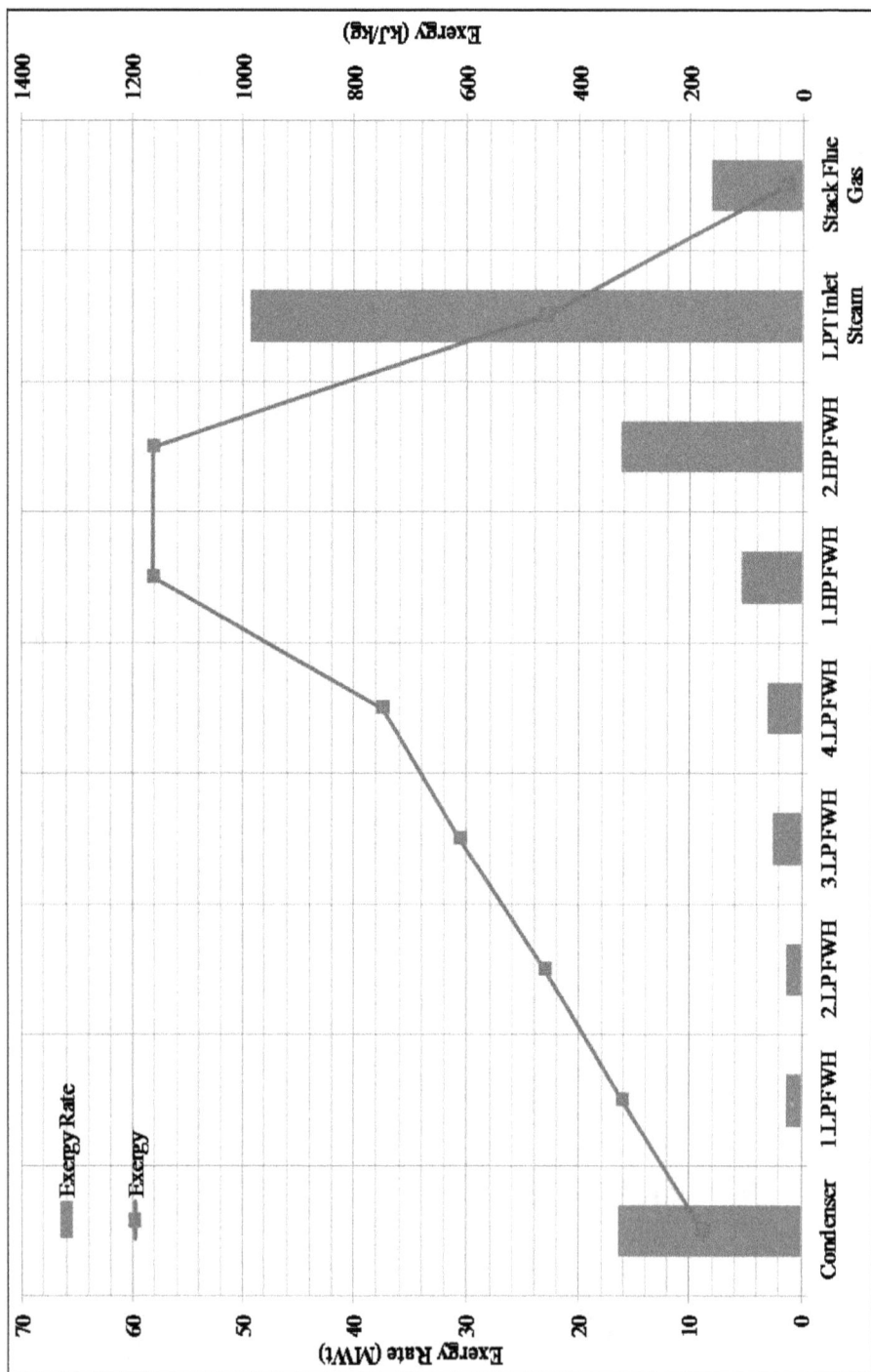

Figure 15.7: TPP Heat Sources Exergy and Exergy Rates

3. The plant power production and thermal efficiency should be minimum affected: After utilization of extracted steam in DH application total exergetic efficiency should increase and exergy losses per exergy output should decrease. First and second law of thermodynamics states that more close extracted steam exergy to the DH water, the less energetic and exergetic losses.

4. Plant's physical conditions should be available for retrofitting: If heat extraction point does not give physical permission and it will give lack of process in the TPP cycle, this source cannot be appropriate for DH usage.

5. The source should be in a potential supplying maximum heat demand and heat supply safety of the city: The sources should be selected with respect to DH system maximum heat demand. According to some problems in system, heat supply safety should be also taken into consideration.

According to these criterias, steam line between intermediate pressure (IPT) turbine and low pressure (LPT) turbine was selected as a steam extraction point for base heat load (Table 15.5). The steam conditions at this point are; 226 °C temperature and 2.3 bar pressure based on nominal load design values. The conceptual design of the DH conversion of TPP for one unit and Soma DH system is given in the Figure 15.8.

Table 15.5: Comparisons of Heat Potentials

Heat Sources	1	2	3	4	5
Condenser	X	X	✔	✔	✔
Stack flue gas	✔	✔	✔	~	X
1. LP FWH	✔	X	✔	X	X
2. LP FWH	✔	X	✔	X	X
3. LP FWH	✔	✔	✔	X	X
4. LP FWH	✔	✔	~	X	X
1. HP FWH	✔	✔	X	X	X
2. HP FWH	✔	✔	X	X	X
LPT inlet	✔	✔	~	✔	✔

In DH systems, supply reliability is also very important, so for this reason the live steam from 1-6 RQ (cold reheat steam) collector (connecting to all six units in TPP) and three hot water boilers are adapted to the conceptual design of Soma City DH system (Figure 15.8).

Steam extraction from crossover pipe between IPT and LPT causes some change in electricity. However, most of the heat is gained from the waste part of condenser waste heat. Thus, it can be said that most of the heat extracted from LPT inlet is waste heat. This case is illustrated in Figure 15.9. The quantity of the waste heat usage is increased as the rate of extracted steam from crossover pipe increases even with small change in electricity production.

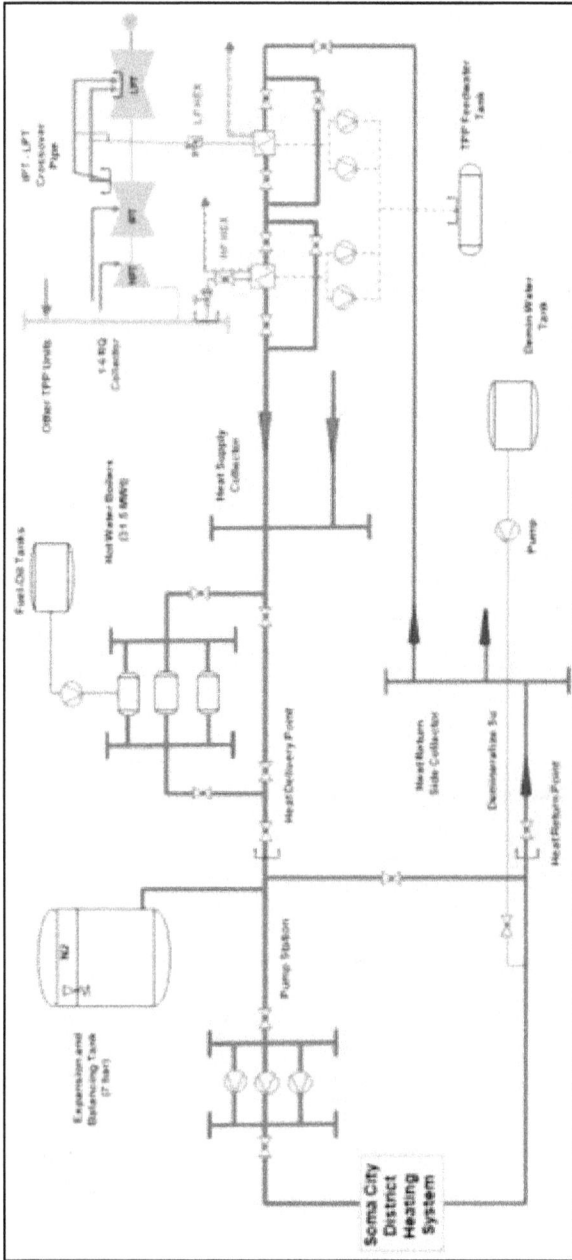

Figure 15.8: Schematic Diagram of Steam Extraction for DH System in Unit Cycle3

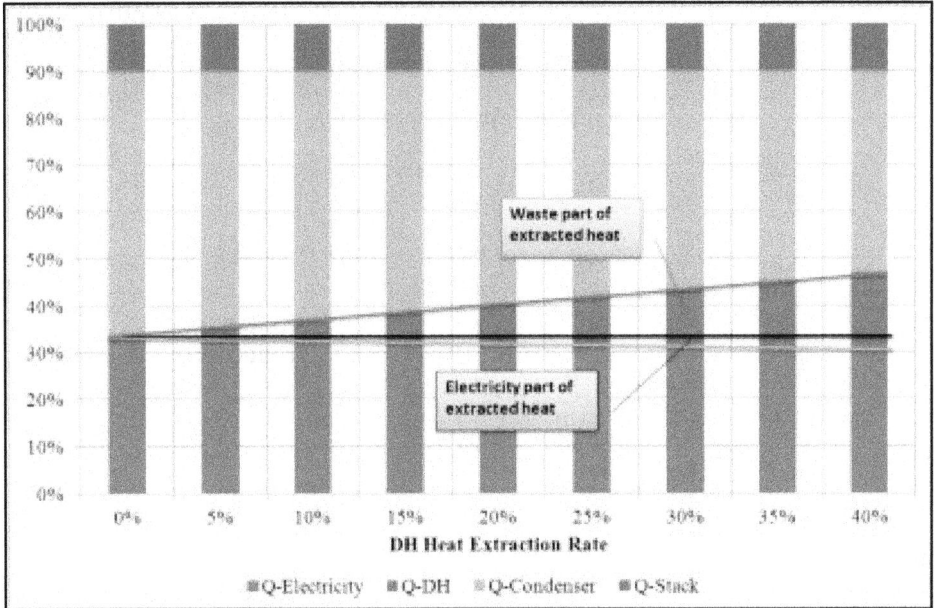

Figure 15.9: Waste Part of Extracted Steam with Extraction Rate

Additionally, according to thermodynamic analysis on TPP cycle with DH adaptation, it is showed that a certain amount of electricity decrease (ΔE) will be as such a steam extraction is applied for DH. It is because of the reduced amount of steam at the crossover pipe to the LPT. These values can be calculated depending on

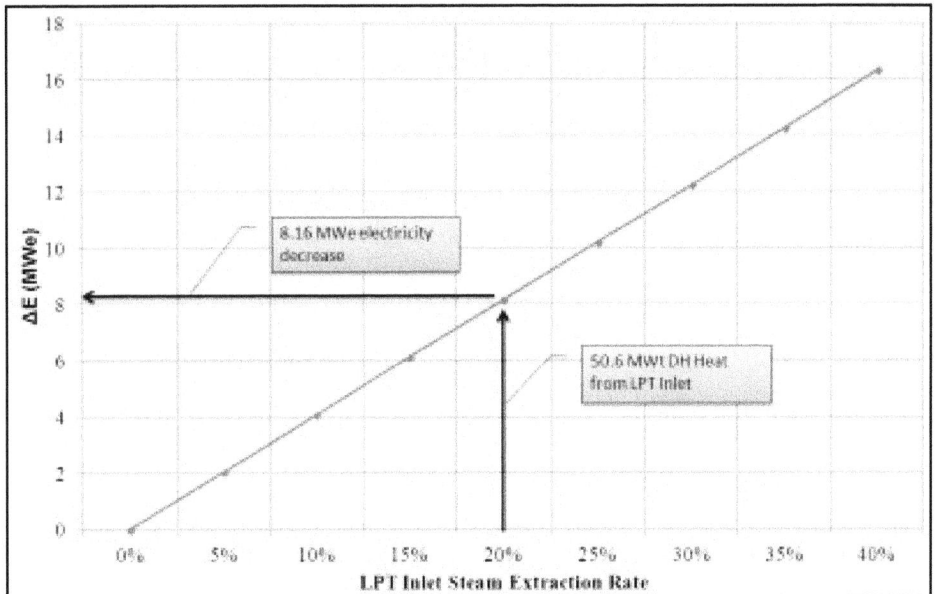

Figure 15.10: Change in Electricity with Steam Extraction from LPT Inlet

the extracted steam rate (R_{DH}) by the equations 9 and 10 and results are given in the Figure 15.10.

$$\Delta E = m_{DH}(h_8 - H_{12})/lpt \qquad [9]$$

$$R_{DH} = \frac{m_{DH}}{m_s} \qquad [10]$$

Here, m_{DH} is extracted steam mass flow rate from LPT inlet. m_8 and h_8 are the total mass steam flow rate and enthalpy at crossover pipe. h_{12} is the enthalpy value at the exit of the LPT. $_{lpt}$ is the efficiency of low pressure turbine. As examined the Figure 10, it is seen that steam extraction changing the electricity production 1 MWe supplies 6.2 MWt heat to the DH system. It means that 5.2 MWt heat is gained from the waste heat that will be released from condenser cooling system.

3. CONCLUSIONS

In this study, energy and exergy analyses of one unit of Soma B TPP have been conducted by developed simulation model. By taking Soma City DH system heat demand and TPP related technical issues into consideration, the heat source for Soma City DH system is selected and conceptual design completed. Summary of results from the study is given as follows:

☆ For energy efficiency, the new TPP should be established both generating electricity and heat for the nearby city. This will increase total efficiency of the usage of primary energy sources

☆ If existing TPPs, generating only electricity, are very close to the cities, the feasibility of the TPP should be carried out in order to benefit from waste heat sources in TPP

☆ For the heat source search in TPP, the energetic and exergetic analyses for waste heat potentials for condenser, stack gases, feed water heaters and low pressure turbine inlet can be performed

☆ Waste stack gas potential is technically feasible with its no effect on TPP performance with low exergy value and its source temperature. On the other hand, this source is very limited (26 per cent) as 64 MWt Soma City DH heat demand is concerned for 8,100 dwellings. Thus, it is a problem for supply reliability. In addition, it has high investment cost as the gas side has low heat transfer coefficient

☆ Condenser has the highest potential waste heat with low exergy value and low costs. However, temperature is not convenient for DH systems. This potential can be evaluated by low temperature heating. Investment costs are high as the total flow rate for heating purposes should be high to provide with DH heat demand

☆ Analyses showed that the best point for DH system heat demand is extraction of LPT inlet steam from the crossover pipe. This kind of applications is also performed in some TPPs abroad. Thus, it is technically feasible. Its potential is high enough to supply heat to approximately 7,000 dwellings. As there are 6 units in Soma B TPP, total heat from this source is adequate for all Soma City for 22,100 dwellings

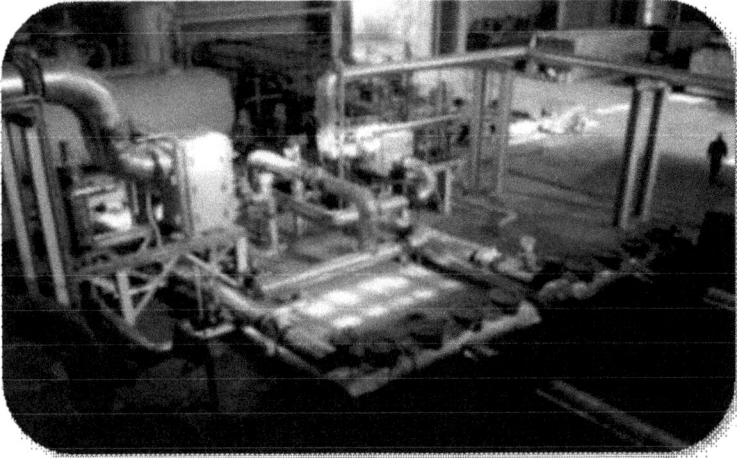

Figure 15.11: A Photo of Technical Converison at Soma B TPP for Cogeneration

☆ As steam is extracted from LPT inlet for DH heat demand, one point change in electricity production corresponds to 5.2 points decrease in waste heat from the condenser. Totally 6.2 point heat is gained. Thus, the extracted heat from this point is mostly waste heat that will be sent to atmosphere if used in LPT

☆ This is basic general model for performance evaluation of exiting TPP, generating only electricity, in order to find sources for heating purposes. Thus, this model can be employed for other conventional TPPs as well

Figure 15.12: Soma City Transmission Pipeline Installations

☆ For the realization of the project, technical conversion of one TPP unit into cogeneration is completed Figure 15.11). Also, Soma City has established 2.5 km transmission line and some extent of distribution lines for DH system (Figure 15.12). At present, TPP supplies heat to 900 dwellings in Soma City for heating and hot tap water supply purposes. DH system continues to disseminate in Soma and it is assumed that it will reach 8,100 dwellings in 3 years

4. ACKNOWLEDGEMENTS

The research activities in this paper were supported by the Scientific and Technological Research Council of Turkey (TUBITAK) (Project no. 105G099). In addition, the author acknowledges the support from by the Electricity Generation of Turkey Co. Inc. (EUAS).

REFERENCES

1. Annual Report, Electricity Generation Co. of Turkey, 2012.

2. 26th BTYK Presentation, 2012.

3. E. Karlopoulos, D. Pekopoulos, E. Kakaras, 2004. District heating systems from lignite fired power plants ten years experience in Greece. http: www.opetchp.net/download7.asp (accessed in 2009)

4. P. Sonne, 2006. Kozani – a success story in district heating world. DBDH 3, p. 14-15.

5. C. Güngör, H.H. Erdem, B. Sahin, I.Teke, A. Dagdas, S.H. Sevilgen, S. Atas, M. Kahraman, G. Gundogdu, *et al.*, 2010. Research and development of methods to benefit from thermal power plant surplus heats and district heating applications (Project no. 105G099). Soma DH system feasibility study (in Turkish).

6. H.H. Erdem, A.V. Akkaya, B. Cetin, A. Dagdas, *et al.*, 2009. Comparative energetic and exergetic performance analyses for coal-fired thermal power plants in Turkey. International Journal of Thermal Sciences, 48 : 2179-2186.

7. H.H. Erdem, A.V. Akkaya, B. Cetin, A. Dagdas, *et al.*, 2010. Thermodynamic analysis of an existing coal-fired power plant for district heating/cooling application. Applied thermal engineering, 30:181-187.

8. J.G.S. Fonseca, P.S. Schneider, 2006. Simulation of a thermal power plant with district heating: comparative results of 5 different codes. Energy, 31:1955-1968.

9. District heating and cooling system using waste steam. Caddet Energy Efficiency Brochure, 1996, JP96.501/5A.D01.

10. V. Verda, R. Borchiellini, M. Cali, 2001. A thermoeconomic approach for analysis of district heating systems. International Journal of Appllied Thermodynamics, 4 (4) p. 183-190.

11. C. Güngör, H.H. Erdem, B. Sahin, I.Teke, A. Dagdas, S.H. Sevilgen, S. Atas, M. Kahraman, G. Gundogdu, *et al.*, 2010. Research and development of methods to benefit from thermal power plant surplus heats and district heating applications

(Project no. 105G099). Potential specification analyses and utilization methods report (in Turkish).

12. C. Güngör, H.H. Erdem, B. Sahin, I.Teke, A. Dagdas, S.H. Sevilgen, S. Atas, M. Kahraman, G. Gundogdu, *et al.*, 2010. Research and development of methods to benefit from thermal power plant surplus heats and district heating applications (Project no. 105G099). Pilot application scope report (in Turkish).

Annexure

Bengaluru Declaration on Sustainable Energy for All

WHILE EXPRESSING deep gratitude to the Society of Energy Engineers and Managers (SEEM) and the Centre for Science and Technology of the Non-Aligned and Other Developing Countries (NAM S&T Centre) for organising the 3rd Triennial International Workshop on 'Sustainable Energy for All: Transforming Commitments to Action', which was held at Bengaluru, India during 22-24 February 2014, as well as to Christ University Faculty of Engineering in Bengaluru for being the gracious host of the event;

HAVING BEEN CONVINCED that the annual global energy sector investments of more than US$1 trillion, much of it going towards traditional and high-carbon energy systems can be directed towards building the clean energy economies of the future, the world needs transition from its current unsustainable energy paradigm to a future substantially complemented by renewable energy sources with high levels of resource efficiency, thus requiring sustainable energy technologies to play a key role in reducing the world's dependence on non-sustainable, fossil-based energy sources;

BY SHARING the experiences in Non Aligned and other developing countries and showcasing country experiences in ensuring universal access to modern energy services, improving energy efficiency and increasing the share of renewable energy in the energy mix;

RECOGNISING that the expansion of sustainable energy technologies such as solar, biomass, wind, hydro, geothermal, ocean resources, low-carbon transportation and the energy storage technologies will make a substantial contribution in reducing the environmental and cost burdens associated with conventional energy supplies, also leading to a decline in the use of fossil fuels and greenhouse gas emissions associated with climate change;

REALISING that the investment in sustainable energy is a smart strategy for growing markets in developing countries, improving competitiveness, increasing energy security and providing greater equity and opportunity,

THE PARTICIPANTS FROM AUSTRALIA, EGYPT, INDIA, INDONESIA, IRAQ, MALAYSIA, MAURITIUS, MYANMAR, NIGERIA, THE PHILIPPINES, SRI LANKA, THAILAND, TURKEY, VIETNAM, ZAMBIA AND ZIMBABWE EXPRESS BELIEF that the deliberations of the workshop on 'Sustainable Energy for All 2014' were a success in sharing of knowledge, facilitating South-South and North-South knowledge transfer, at the end of which the following recommendations were made, including some action items, for adoption by the participating countries in formulating their policies and action plans to improve access to reliable, affordable, economically viable, socially acceptable and environmentally sound energy services and resources for sustainable development, and

UNANIMOUSLY RESOLVE that all developing countries are strongly encouraged to:

☆ Exhibit a strong political commitment, endorse the 'Sustainable Energy for All' vision, and develop unilateral, bilateral, multilateral and regional action plans to ensure universal energy access, enhancing energy efficiency and scaling-up renewable energy;

☆ Identify existing gaps in policy, regulatory framework and technology arena and accordingly develop appropriate schemes to promote renewable energy, energy efficiency and energy access;

☆ Develop suitable protocols for energy auditing and benchmarking energy consumption, and prepare appropriate action plans to reduce energy intensity and carbon footprint;

☆ Ensure that "modern energy services" do not necessarily mean new equipment which directly or indirectly increase electricity consumption, but rather it could be the improved versions of traditional systems and/or new energy efficient innovative products;

☆ Strengthen mechanisms for institution-industry collaboration for R&D, technology transfer and human resource development and to provide appropriate facilitation of sustainable energy market development;

☆ Establish a coordination mechanism to engage all the concerned government departments including energy, environment, science & technology, economic affairs, planning, health, education, water, rural and urban development, industry, agriculture et cetera, while developing policies, regulations and financing mechanisms for clean energy services due to its across-the-board impact;

☆ Keep the interplay between energy security and food security in mind since the issue of energy security compromising the capability of countries to meet their food needs has come up as an emerging concern leading to a definite shift from first-generation bio-fuels based on food crops to next-generation bio-fuels made from inedible feedstock;

☆ Promote participation of business and financing establishments in sustainable energy programmes, considering the need to scale up the level of action and investment;

☆ Engage civil society organisations and NGOs and empower beneficiary communities especially women and youth, given the extent of action and advocacy needed;

☆ Educate all stakeholders to practice energy modesty and use energy rationally in the context of the energy rebound effect;

☆ Strengthen information and communication activities for effective outreach and campaigns, soliciting active involvement of print, visual, electronic and social media to bring about positive attitudinal change towards adoption of sustainable energy systems.

This declaration calls upon governments, business and civil society to reiterate their obligation to promote a healthy society that encourages the wellbeing of all people and ensures a clean, green, safe planet to our coming generations.

Participants appreciated the proposal from SEEM to establish a joint Fellowship scheme on Sustainable Energy with the NAM S&T Centre, subject to necessary approvals and availability of funds.

THUS RESOLVED AND ADOPTED ON THE 24ᵗʰ FEBRUARY 2014 AT BENGALURU, INDIA.